Malaria: Genetic and Evolutionary Aspects

Emerging Infectious Diseases of the 21st Century

Series Editor: I. W. Fong
Professor of Medicine, University of Toronto
Head of Infectious Diseases, St. Michael's Hospital

Recent volumes in this series:

INFECTIONS AND THE CARDIOVASCULAR SYSTEM: New Perspectives
Edited by I. W. Fong

REEMERGENCE OF ESTABLISHED PATHOGENS IN THE 21ST CENTURY
Edited by I. W. Fong and Karl Drlica

BIOTERRORISM AND INFECTIOUS AGENTS: A New Dilemma for the 21st Century
Edited by I. W. Fong and Kenneth Alibek

MALARIA: Genetic and Evolutionary Aspects
Edited by Krishna R. Dronamraju and Paolo Arese

A Continuation Order Plan is available for this series. A continuation order will bring delivery of each new volume immediately upon publication. Volumes are billed only upon actual shipment. For further information, please contact the publisher.

Malaria: Genetic and Evolutionary Aspects

KRISHNA R. DRONAMRAJU
AND
PAOLO ARESE

Krishna R. Dronamraju
President, Foundation for
Genetic Research
P.O. Box 27701-0
Houston, Texas 77227
USA
KDronamraj@aol.com

Paolo Arese
Department of Genetics, Biology
and Biochemistry
University of Torino Medical School
Via Santena 5 bis, 10126 Torino
Italy
paolo.arese@unito.it

ISBN 978-1-4419-2102-4 e-ISBN 978-0-387-28295-4

Softcover reprint of the hardcover 1st edition 2006

9 8 7 6 5 4 3 2 1

springeronline.com

Contents

5. Resistance to Antimalarial Drugs: Parasite and Host Genetic Factors

Rajeev K. Mehlotra and Peter A. Zimmerman

6. Evolutionary Origins of Human Malaria Parasites

Stephen M. Rich and Francisco J. Ayala

7. Vector Genetics in Malaria Control

V.P. Sharma

Preface

This book was originally conceived at a conference at the University of Turin in Italy. The conference was organized to examine the so-called "Malaria Hypothesis", that is to say, the higher fitness of thalassemia heterozygotes in a malarial environment, and to pay tribute to the proponent of that hypothesis, J.B.S. Haldane. Contributors to this book examine certain genetic and evolutionary aspects of malaria which is a major killer of human populations, especially in Africa and Asia.

There were attempts to discredit Haldane's contribution from two directions: (a) it has been suggested that the "Malaria Hypothesis" was known long before Haldane and that there was nothing original about his idea (Lederberg 1999), and that (b) the hypothesis of heterozygote superiority was first suggested by the Italian biologist Giuseppe Montalenti who communicated his idea to Haldane (Allison 2004). Surely, both cannot be right. In fact, the evidence presented in this book clearly indicates that both are wrong.

Haldane's malaria hypothesis has stimulated a great deal of research on the genetic, evolutionary and epidemiological aspects of malaria during the last 50 years. It has opened up a whole new chapter in the study of infectious diseases. It deserves serious consideration.

For helpful discussions we thank Lucio Luzzatto, Alberto Piazza, Guido Modiano and David Roberts.

Krishna R. Dronamraju
Paolo Arese

Preface

Contributors

Paolo Arese
Department of Genetics, Biology and Biochemistry
University of Torino Medical School
Via Santena 5 bis
10126 Torino
Italy

Francisco J. Ayala
Department of Ecology and Evolutionary Biology
University of California
Irvine, California 92697
USA

Kodjo Ayi
Department of Genetics, Biology and Biochemistry
University of Torino Medical School
Via Santena 5 bis
10126 Torino
Italy

Stefano Canali
University of Cassino
Via Passo Cento Croci
1-00048 Nettuno (Rome)
Italy

Gilberto Corbellini
Section of History of Medicine
"La Sapienza" University of Rome
Viale dell'Universita 34/a
00185 Rome
Italy

Krishna R. Dronamraju
President, Foundation for Genetic Research
P.O. Box 27701-0
Houston, Texas 77227
USA

Rajeev K. Mehlotra
Center for Global Health and Diseases
Case Western Reserve University
School of Medicine
Wolstein Research Building, # 4125
2103 Cornell Road
Cleveland, Ohio 44106-7286
USA

Stephen M. Rich
Department of Plant, Soil and Insect Sciences
University of Massachusetts
Amherst, MA 01002
USA

V.P. Sharma
Centre for Rural Development and Technology (CRDT)
Indian Institute of Technology (IIT)
Hauz Khas,
New Delhi 110016
India

Aleksei Skorokhod
Department of Genetics, Biology and Biochemistry
University of Torino Medical School
Via Santena 5 bis
10126 Torino
Italy

Franco Turrini
Department of Genetics, Biology and Biochemistry
University of Torino Medical School
Via Santena 5 bis
10126 Torino
Italy

Peter A. Zimmerman
Center for Global Health and Diseases
Case Western Reserve University
School of Medicine
Wolstein Research Building, # 4125
2103 Cornell Road
Cleveland, Ohio 44106-7286
USA

Introduction

Krishna R. Dronamraju

The global incidence of malaria and its fatal consequences continue to be one of the worst catastrophes ever faced by mankind. With over 3 million deaths per year that are attributable to the attacks by the malaria parasite, it remains the foremost killer among all diseases, comparable only to the deaths caused by the equally disastrous disease, AIDS, which was reported to have caused 3.1 million deaths in 2004 (UNAIDS/WHO AIDS Update December 2004).

However, by utilizing a combination of epidemiological, geographical, and demographic data on the occurrence of malaria, Snow *et al.* (2005) showed that there is reason to believe that these passive statistics may well indicate gross underestimates. The total number of deaths that are occurring each year from malaria (and also perhaps from AIDS and other infectious diseases) may be much higher than indicated by these estimates.

The global distribution of clinical attacks attributable to the parasite, causing the common malaria, *Plasmodium falciparum*, has been recently shown to be grossly underestimated in previous reports. Snow *et al.* (2005), in a remarkable paper, estimated that there were 515 (range 300–660) million episodes of clinical *P. falciparum* malaria in 2002. Their estimate is approximately 50% higher then those reported by the World Health Organization (WHO) and 200% higher for areas outside Africa. This difference was explained by Snow *et al.* (2005) as the result of the dependence on national statistics derived from passive detection of cases for the WHO's present global disease estimates outside Africa. Similar passive data, when compared with survey reports of data on active case detection in the same areas, in various countries, demonstrated that the magnitude of underreporting by passive detection ranged from a 3-fold difference in Brazil to a 1000-fold difference in Pakistan.

Krishna R. Dronamraju • Foundation for Genetic Research, P.O. Box 27701-0, Houston, Texas 77227.

Malaria: Genetic and Evolutionary Aspects, edited by Krishna R. Dronamraju and Paolo Arese, Springer, New York, 2006.

1. The Haldane Hypothesis

A new chapter on the evolutionary implications of the prevalence of infectious diseases in human populations was opened by the biologist J.B.S. Haldane (1932) who suggested that one of the principal agents of natural selection during the course of evolution is immunity to disease (Dronamraju, 2004). Later, in two seminal papers that were delivered in 1948 and 1949, Haldane (1949 a,b) developed his theory succinctly. These are reproduced in this book (pp.169–187). In his first paper, an address to the Eighth International Congress of Genetics in Stockholm, Haldane (1949a) discussed the possibility that individuals who are heterozygous for thalassemia in the Mediterranean countries may be more resistant to malarial infection. He was impressed by the co-occurrence of both thalassemia and malaria in countries of the Mediterranean region. He argued that the corpuscles of the anemic heterozygotes may well be more resistant to attacks by the sporozoa, which cause malaria than those of either homozygote. His comment was in response to a statement by Neel and Valentine (1947) who believed that the thalassemia heterozygote is less fit than normal and the mutation rate may well exceed 4×10^{-4}. Haldane (1949a) noted that if the heterozygote had an increased fitness of only 2%, this would account for the incidence without invoking any mutation at all. He added that an increased fitness of the heterozygote was also found in the case of several lethal and sublethal genes in *Drosophila* and *Zea*.

Later, at a symposium in Milan, Haldane (1949b) developed his theory in greater detail, extending it to infectious diseases in general. He wrote: ". . . the struggle against disease, and particularly infectious disease, has been a very important evolutionary agent, and some of its results have been rather unlike those of the struggle against natural forces, hunger, and predators, or with members of the same species." Haldane did not refer to thalassemia or malaria in his paper, but during the discussion the Italian biologist Giuseppe Montalenti referred to the greater fitness of thalassemic heterozygotes in the malarial regions, which, he said, was verbally communicated to him by Haldane. Thus, it is clear that Haldane was the first to formulate the "malaria hypothesis."

Lederberg (1999), stated that genetic resistance to infectious disease in crops, such as wheat, was known long before Haldane and that Haldane (1949b) did not refer directly to the relationship between malaria and thalassemia. However, Lederberg (1999) did not mention the earlier paper of Haldane (1949a) where Haldane clearly discussed the role of heterozygosity in connection with resistance to malaria. Allison (2004), suggested the possibility that it was Montalenti who communicated the hypothesis to Haldane. However, this possibility was also ruled out for two reasons. (a) It was Montalenti himself who stated that Haldane communicated the idea to him.

(b) Haldane had already discussed his hypothesis at the International Genetics Congress in Stockholm in 1948, long before the Milan symposium.

It is hardly surprising that Haldane came up with the idea of a relationship between malaria and thalassemia. Throughout his career, from 1912 onwards, Haldane was noted for his broad vision and far reaching intellect, which he displayed in several disciplines to which he made fundamental contributions. He possessed the rare gift of seeing connections between diverse observations and phenomena in multiple disciplines, which others had missed. These disciplines include physiology, genetics, biochemistry, biometry, statistics, and cosmology, to name a few. His mathematical theory of natural selection was one of the foundations of population genetics (Dronamraju, 1990). Through long and consistent mathematical contributions over 50 years and sweeping generalizations in evolutionary biology, which encompassed his vast knowledge of multiple disciplines, Haldane's broad vision and insight were most influential in shaping the course of evolutionary biology. Unlike the work of his contemporaries, R.A. Fisher and Sewall Wright, Haldane's contributions to biology were much more extensive, covering many aspects of human–medical genetics, population genetics, demographic genetics, immunogenetics, behavior genetics, cytogenetics, as well as other subjects including cybernetics, philosophy, logic, astronomy, and psychology. No other individual matched his versatility. Indeed, it is clear from today's perspective that the "malaria hypothesis" was yet another idea of Haldane, which was one of a long series of brilliant ideas, hypotheses, and theories that transformed biology, especially evolutionary biology, in a profound manner. It is entirely consistent with Haldane's approach to problems in biology. His numerous books, research papers, and popular essays often contained valuable ideas, which are too numerous to mention here. Some of these are found in the most unlikely places, such as the The Rationalist Annual or *The Daily Worker*, and were mentioned so casually that it is easy to overlook them (Dronamraju, 1985).

Haldane's "malaria hypothesis" was recently examined and reviewed by Weatherall (2004). It is an interesting fact that Haldane's hypothesis was first confirmed in the sickle-cell anemia field rather than thalassemia. These early studies on thalassemia were reviewed in detail by Weatherall and Clegg (2001). Recent research has provided stronger numerical data regarding the protective effect of the sickle-cell trait against *P. falciparum* malaria. Sickle-cell carriers enjoy almost 80% protection against the severe complications of malaria, especially cerebral and profound malaria (Hill *et al.*, 1991). Research in West Africa by Modiano *et al.* (2001) has confirmed that the relatively high frequencies of Hb C have been maintained by resistance to *P. falciparum*. However, there appears to be evidence for both heterozygote and homozygote protection and the authors suggested that this could be an example of a transient polymorphism, mainly because of the perceived lack of any clinical or hematological changes in Hb C homozygote.

Research involving both population and case-control studies has provided strong evidence that the high frequency of the milder varieties of alpha-thalassemia is related to protection against *P. falciparum* malaria (reviewed by Weatherall and Clegg, 2002). There is strong evidence of protection of both alpha$^+$ thalassemia homozygotes and heterozygotes against *P. falciparum*. Molecular analyses of beta-globin genes in thalassemic and nonthalassemic individuals in different populations provided some indirect evidence that the high frequency of beta-thalassemia is also indicative of heterozygote protection against malaria. However, this observation needs further confirmation by the application of case-control studies, which have proved to be successful with respect to the sickle-cell and alpha-thalassemia genes.

Although Haldane (1949a) offered a possible explanation, it is clear that it was overly simplistic. Recent investigations with *in vitro* cultures of malarial parasites have yielded some interesting results. Analyses of rates of invasion and growth of parasites in normal and thalassemic red cells, as well as in the red cells of carriers of Hb S and Hb C have been carried out by various investigators (Nagel, 2001; Weatherall and Clegg, 2001). Weatherall (2004) summed up the results. It appears that both *P. vivax* and *P. falciparum* have a predilection for invading younger red cells (Pavsol *et al.*, 1980). Some abnormalities of invasion and growth have been found in the red cells of individuals with more severe forms of thalassemia. However, no consistent abnormalities were found in the milder forms. Consistent reduction in parasite development has been found in red cells from those with the sickle-cell trait, provided the cells are subjected to hypoxic conditions. In other studies, it has been suggested that protection against malaria may involve an altered red-cell membrane band 3 protein, which may be a target for enhanced antibody binding in thalassemic red cells (Williams *et al.*, 2002).

In vivo studies have also yielded some interesting findings. It has been suggested that early susceptibility to *P. vivax* may be acting as a natural vaccine by inducing cross-species protection against *P. falciparum* (Willams *et al.*, 1996). There has also been a suggestion from case-control studies that thalassemia clearly protects against malaria and that it may be associated with protection against other infectious diseases as well (Allen *et al.*, 1997).

2. Glucose-6-phosphate dehydrogenase deficiency and malarial resistance

Tishkoff and Verrelli (2004) recently reviewed the evidence concerning the relationship between malarial resistance and another polymorphism, G6PD (glucose-6-phosphate dehydrogenase) deficiency. Glucose-6-phosphate dehydrogenase is an important "housekeeping" enzyme in the glycolytic pathway for glucose metabolism. Glucose-6-phosphate dehydrogenase deficiency is a genetic disorder, which results from mutations within the G6PD gene, located on the telometric region of the long arm of the X chromosome. It is the most common enzymopathy of humans, which affects approximately 400 million people worldwide (Vulliamy *et al.*, 1992). Glucose-6-phosphate dehydrogenase deficiency has been found to be associated with many clinical disorders, including hemolytic anemia, neonatal jaundice, and several cardiovascular diseases, which may be triggered by certain drugs, such as the antimalarial drug primaquine or certain foods or infections. For example, ingestion of fava beans by individuals with the common Mediterranean G6PD variant has been known to trigger severe hemolytic anemia and death, resulting in the disease commonly known as "favism" (Vuilliamy *et al.*, 1992; Beutler, 1993). The distribution and frequency of G6PD deficiency are positively correlated with regions where malaria is endemic—either now or in the past (Siniscalco *et al.*, 1961). There is abundant evidence confirming that the individuals with G6PD deficiency have lower *P. falciparum* parasite loads than controls (Gilles *et al.*, 1967). Furthermore, in heterozygotes for G6PD deficiency, more parasites are present in cells with normal enzyme activity than in cells with deficient enzyme activity (Luzzatto *et al.*, 1969). Parasite growth is inhibited in the first few cycles of infection in G6PD deficient cells. A large case-control study involving more than 2000 African children demonstrated that the most common African form of G6PD deficiency is associated with a 46–58% reduction in risk of severe malaria for both female heterozygotes and male hemizygotes (Ruwende *et al.*, 1995). The precise mechanism by which G6PD deficiency confers resistance against malaria is not known. It has been suggested that red cells deficient in G6PD may be under increased oxidative stress, which may create a toxic environment for the *P. falciparum* parasites (Miller, 1994). Finally, the G6PD-deficient red blood cells (RBC) that contain parasites at the ring stage of infection and have a greatly impaired antioxidant defense are more likely to undergo phagocytosis, leading to a reduced number of red cells containing mature parasites.

3. Phagocytosis of Ring Forms

Arese *et al.* (see pp. 25–54 of this book) have discussed the enhanced phagocytosis of early parasite forms (ring forms) as a model of protection for widespread RBC mutations. Their model suggests that only phagocytosis of ring-parasitized mutant RBCs is selectively enhanced, while phagocytosis of trophozoites, although very high, is quite similar in normal and mutant cells. This model of resistance, which involves an enhanced phagocytosis of ring-parasitized mutant RBC, was earlier proposed by Cappodoro *et al.* (1998) for G6PD deficiency and was later extended to Hb AS, beta-thalassemia trait, and Hb H by Ayi *et al.* (*Blood*, in press). It has been noted that several protective mutations have phenotypic similarities, although differing in the molecular nature of the underlying defect (Destro-Bisol *et al.*, 1996). As expected, higher levels of bound antibodies and more intense phagocytosis have been observed in parasitized thalassemic and other mutant RBC. It appears that the mechanism of underlying resistance is due to the enhanced removal of parasitized mutant RBC by the host's immune system.

4. Evolutionary Considerations: Malaria's Eve Hypothesis

Rich and Ayala (1998, 2004, and in the present volume, see pp. 125–146) have discussed the evolutionary origins of human malaria parasites. Some interesting facts are as follows. The four human parasites, *P. falciparum*, *P. ovale*, *P. malariae*, and *P. vivax* are very remotely related to each other, which imply that the evolutionary divergence of these four human parasites greatly predates the origin of the hominids. The evidence indicates that their parasitic associations with humans are phylogenetically independent, which is further confirmed by the diversity of their physiological and epidemiological characteristics. *P. falciparum* is more closely related to *P. reichnowi*, the chimpanzee parasite, than to any other *Plasmodium* species mentioned earlier. The time of divergence between *falciparum* and *reichnowi* has been estimated to be about 8–11 million years, which is in agreement with the time of divergence between the humans and the chimpanzees. In addition, there is evidence to indicate parasite transfers between human and monkey hosts. Evidence favoring the transfer from monkeys to humans and *vice versa* has been summarized by Rich and Ayala (2004 and this volume, pp. 125–146). Lateral transmission of *Plasmodium* parasites from monkey hosts to humans is known for *P. simium*, *P. brasilianum*, *P. cynomolgi*, *P. knowlesi*, and also possibly *P. simiovale*. Transmission from humans to monkeys has been achieved experimentally and may occur in nature as well. Host-shifts have been noted among avian and reptilian malaria parasites (see Rich and Ayala, this volume).

Rich and Ayala (1998) proposed that the extant world populations of *P. falciparum* are of recent origin, referring to it as the "Malaria's Eve hypothesis." Although this has been a source of some contention in recent years, the work of Conway *et al.* (2000) provided further evidence in support of this hypothesis on the basis of an analysis of the *P. falciparum* mitochondrial genome (however, also see Volkman *et al.*, 2001; Hughes and Verra, 2001; Rich and Ayala, 2004).

Coluzzi (1994, 1999) argued that speciation events that may be related to the demographic and climatic changes in Africa, the Middle East, and the Mediterranean region had facilitated the rapid evolution of highly anthropophilic and effective *falciparum* vectors. Della Torre *et al.* (2002), (from Coluzzi's group) studied the molecular affinities among the sibling species of the *Anopheles gambiae* complex, which are morphologically indistinguishable but exhibit distinct genetic and eco-ethological differences which, in turn, are related to differences in their ability to transmit malaria. *A. gambiae* shows extreme genetic heterogeneity, which is revealed by the traditional study of chromosomal inversions but also by recent studies of molecular markers, such as X-linked ribosomal DNA (rDNA). The very recent divergence of the molecular forms of the *A. gambiae* complex and the likelihood that only a few genes are involved in reproductive isolation and ecological diversification means that the entire *A. gambiae* genome needs to be screened to identify differences in gene sequences and gene expression between incipient species (Della Torre *et al.*, 2002). This situation offers a unique opportunity to study the evolutionary strategies of a vector, which has survived and evolved over thousands of years. It is necessary to understand this evolutionary process if we are to devise methods to control this vector, which continues to cause much destruction to human society.

Among other evolutionary considerations, Modiano *et al.* (2001) suggested that Hb C provides protection against clinical *P. falciparum* malaria in both the heterozygous and homozygous state. The estimated reduction in the relative risk of clinical malaria associated with CC homozygosity (93%) is stronger than that of AC heterozygosity (29%) and similar to that of the Hb AS genotype (73%). Based on these findings as well as other work on pathological aspects, they have suggested that in the long term and in the absence of malaria control, Hb C would replace Hb S in central West Africa.

Alphey *et al.* (2002) suggested that genetically modified organisms (GMOs) could be used to engineer mosquitoes with an altered phenotype that would be introduced into a population and spread rapidly. These strategies would target the malaria parasite rather than the mosquito itself. Another possibility is to follow a traditional method, such as the release of insects carrying dominant lethals. However, there is much emphasis on testing the safety and efficacy of any novel engineering methods in cultures and in animal models before initiating the clinical trials.

5. Malaria Vaccines

As the parasite, *P. falciparum*, has developed resistance to the mainstays (e.g., chloroquine and sulfadoxine-pyrimethamine), even the modest goal of the "Roll Back Malaria" program of the WHO to cut the death rate by half by 2010 is unlikely to be realized by using conventional methods. Other approaches to malaria eradication are being explored. Since the discovery by Chinese scientists (in the 1970s) that an active ingredient, artimisinin, found in the Chinese shrub, *Artimisia annua,* kills malarial parasites, several new chemical derivatives have been developed. These were found to be quite effective, curing more than 90% of the patients within a few days, with no adverse effects. However, there are two problems. The plant grows mainly in China and Vietnam and its availability for the drug development has not been steady or reliable (although new attempts are being made to grow it in India). The second problem is the cost of treatment—at $ 2.40 per course—is prohibitively expensive for many countries of Africa and Asia where it is urgently needed. The Gates Foundation has been supporting research at the University of California to genetically engineer *Escherichia coli* to make them produce terpenoids, a class of molecules that includes artemisininin. However, it remains to be seen whether the eventual cost of any drug thus produced will be affordable for treating the millions of afflicted people in the developing countries. Another alternative is a compound called OZ277 (or OZ), which has been tested at the University of Nebraska as an antimalarial *in vitro* and in animals. An Indian pharmaceutical company, Ranbaxy, is developing it further, hoping to market it at an affordable price. Even if the price is affordable, drugs can always lose their efficacy sooner or later (sooner, it seems) and the search for newer, safer, and affordable drugs goes on forever.

An obvious alternative is the development of attenuated whole-organism vaccines, which has proved extremely difficult. However, Mueller *et al.* (2005) recently showed that mice immunized with genetically attenuated sporozoites of the rodent malaria parasite *Plasmodium berghei* are completely protected against challenge with wild-type sporozoites. Malaria is transmitted when a female anopheline mosquito injects sporozoites into the host blood. Infection progresses as the sporozoites migrate to the liver, invading hepatocytes and eventually a liver cell where a sporozoite differentiates and multiples, releasing thousands of pathogenic merozoites into the bloodstream. Preventing the sporozoites from infecting a hepatocyte is a prime target for developing malaria vaccines because they can be completely eliminated by sterilizing immune responses, thus preventing malarial infection. Even a single surviving sporozoite can cause a lethal blood-stage infection. Hence, any method devised at this stage must be 100% effective. Other recent developments are also helpful. The recently completed *Plasmodium* genome sequences may facilitate the development of live-attenuated parasites by more accurate genetic interventions. Using expression profiling, the same group of investigators previously identified *Plasmodium* genes that are specifically expressed during the preerythrocyte part of the parasite life cycle. They observed that a number of preerythrocyte genes named *UIS* also undergo upregulation in sporozoites when they gain infectivity for the mammalian host. Hence, the authors concluded that inactivation of *UIS* genes for which expression is restricted to preerythrocyte stages might lead to attenuation of the liver-stage parasite. Their investigation focused on a gene called *UIS3*, which encodes a small conserved transmembrane protein, and is expressed in infectious sporozoites as well as after sporozoite infection of livers *in vivo*. The *UIS3* genes in rodent and human malaria parasites show 34% amino acid sequence identity.

Using gene disruption, Mueller *et al.* (2005) obtained clonal parasite lines designated *uis*(–), which contained the predicted locus deletion. The *uis*(–) parasites showed normal asexual blood-stage growth and normal transmission to the *Anopheles* mosquito vector. The *uis*(–) sporozoites developed normally and infected the salivary glands at wild-type efficiencies. They showed typical gliding motility and host-cell invasion capacity of cultured hepatoma cells at levels comparable to wild-type parasites. However, in contrast to wild-type sporozoites, which are capable of invading hepatocytes and generate merozoites within 48 h, the *uis*(–) parasites showed a severe defect in their ability to complete transformation. These *in vitro* observations were followed up by testing to see if *uis3*(–) sporozoites had lost their capacity to progress through liver-stage development and cause blood-stage infections *in vivo*. These experiments failed to induce blood-stage parasitemia in young Sprague–Dawley rats, which are highly susceptible to *P. berghei* sporozoite infections.

Mueller *et al.* (2005) identified a *UIS3* orthologue in the genome of the human malaria parasite, *P. falciparum*. Indeed, these efforts could lead to the creation of a genetically attenuated *uis3*(–) human parasite that can be tested as a vaccine in human-sporozoite challenge models, eventually leading to the development of a genetically attenuated, protective whole-organism malaria vaccine. Furthermore, the availability of attenuated blood-stage parasites indicates that an attenuated whole-organism approach to malaria vaccination may be developed by focusing attention also upon other stages of the parasite's life cycle. Similar approaches may be explored for other parasite-caused infectious diseases.

Genetically modified parasite sporozoites should be safer than those generated by other methods, such as irradiated sporozoites, provided they are constructed by a replacement (double crossover) strategy so that the possibility of genetic reversion associated with an insertion (single crossover) can be avoided (Menard, 2005). Although such efforts will undoubtedly continue in search of malaria control and eradication, it is also realized that methods to produce vaccines and engineered vectors take much time to develop, and the repeated use of drugs leads to drug-resistant vectors. In the meantime, the number of deaths due to malaria is increasing exponentially, particularly in Africa.

References

Allen, S.J. *et al.* (1997). Alpha+ thalassemia protects children against disease due to malaria and other infections. *PNAS*, **94,** 736–741.

Allison, A.C. (2004). Two lessons from the interface of genetics and medicine. *Genetics*, **166,** 1591–1599.

Alphey, L. *et al.* (2002). Malaria control with genetically manipulated insect vectors. *Science*, **298,** 119–121.

Beutler, E. (1993). Study of glucose-6-phosphate-dehydrogenase—history and molecularbiology. *Am. J. Hematol.*, **42,** 53–58.

Ayi, F. Turrini, A. Piga, and P. Arese (1966). Enhanced phagocytosis of ring-parasitized mutant erythrocytes: a common mechanism that may explain protection against *falciparum* malaria in sickle trait and beta-thalassemia trait. *Blood*, **104**, 3364 - 3371.

Cappodoro, M. *et al.* (1998). Early phagocytosis of glucose-6-phosphate dehydrogenase (G6PD)-deficient erythrocytes parasitized by *Plasmodium falciparum* may explain malaria protection in G6PD deficiency. *Blood*, **92,** 2527–2534.

Coluzzi, M. (1994). Malaria and the Afro-tropical ecosystems impact of man-made environmental changes. *Parassitologia*, **36,** 223–227.

Coluzzi, M. (1999). The clay feet of the malaria giant and its African roots: hypotheses and inferences about origin, spread and control of *Plasmodium falciparum. Parassitologia*, **41,** 277–283.

Conway, D.J., Fanello, C., Lloyd, J.M., Al. Joubri, B.M., Baloch, A.H., Somanath, S.D. Roper. C., Odoula, A.M.J., Mulder, B., Povoa, M.M., Singh, B., and Thomas, A,W. (2000) Origin of *Plasmodium falciparum* malaria is traced by mitochondrial DNA. *Mol. Biochem. Parasitol.*, **111,** 163–171.

Della Torre, A. *et al.* (2002). Speciation within *Anopheles gambiae*—the glass is half full. *Science*, **298,** 115–117.

Dronamraju, K.R. (1985). Haldane: The Life and Work of JBS Haldane with Special Reference to India. Aberdeen University Press, Aberdeen, UK.

Dronamraju, K.R. (ed) (1990). Selected Genetic Papers of J.B.S. Haldane. Garland Publishing Co., New York.

Dronamraju, K.R. (ed) (2004). Infectious Disease and Host-Pathogen Evolution. Cambridge University Press, New York.

Gilles, H.M. *et al.* (1967). Glucose-6-phosphate dehydrogenase deficiency, sickling, and malaria in African children in South Western Nigeria. *Lancet*, **1,** 138–140.

Haldane, J.B.S. (1932). The Causes of Evolution. Longmans, Green & Co., London (reprinted by Princeton University Press, 1990).

Haldane, J.B.S. (1949a). The rate of mutation of human genes. In: Proceedings of the Eighth International Congress of Genetics, *Hereditas*, **35,** 267–273.

Haldane, J.B.S. (1949b). Disease and evolution. *La Ricerca Scientifica*, **19,** 2–11.

Hill, A.V.S. *et al.* (1991). Common west African HLA antigens are associated with protection from severe malaria. *Nature*, **352,** 595–600.

Hughes, A.L. and Verra, F. (2001). Very large long-term effective population size in the virulent human malaria parasite *Plasmodium falciparum. Proc. R. Soc. Lond. B, Biol. Sci.*, **268,** 1855–1860.

Joint United Nations Programme on HIV/AIDS (UNAIDS) and World Health Organization (WHO), WHO, Geneva, 2004.

Lederberg, J. (1999). J.B.S. Haldane (1949) on infectious disease and evolution. *Genetics*, **153,** 1–3.

Menard, R. (2005). Medicine: Knockout malaria vaccine? *Nature*, **433,** 113–114.

Miller, L.H. (1994). Impact of malaria on genetic polymorphism and genetic diseases in Africans and African Americans. *PNAS*, **91,** 2415–2419.

Modiano, D. *et al.* (2001). Hemoglobin C protects against clinical *Plasmodium falciparum* malaria. *Nature,* **414,** 305–308.

Mueller, A.K. *et al.* (2005). Genetically modified Plasmodium parasites as a protective experimental malaria vaccine. *Nature*, **433,** 164–167.

Neel, J.V. and Valentine, W.N. (1947). Further studies on the genetics of thalassemia. *Genetics*, **32,** 38–63.

Nagel, R.L. (2001). Malaria and hemoglobinopathies. In: (M.H. Steinberg, B.G. Forget, D.R. Higgs, and R.L. Nagel (eds) *Disorders of Hemoglobin*. Cambridge, Cambridge University Press, UK, pp. 832–860.

Odoula, A.M.J., Mulder, B. Povoa, M.M., Singh, B., and Thomas, A.W. (2000). Origin of *Plasmodium. falciparum* malaria is traced by mitochondrial DNA. *Mol Biochem Parasitol*, **111,** 163–171.

Pavsol, G., Weatherall, D.J., and Wilson, R.J. (1980). The increased susceptibility of young red cells to invasion by the malarial parasite *Plasmodium falciparum. Br. J. Hematology*, **45,** 285–295.

Rich, S.M. and Ayala, F.J. (1998). The recent origin of allelic variation in antigenic determinants of *Plasmodium falciparum. Genetics*, **150,** 515–517.

Rich, S.M. and Ayala, F.J. (2004). Evolutionary genetics of *Plasmodium falciparum*, the agent of malignant malaria. In: (K.R. Dronamraju, ed) *Infectious Disease and Host-Pathogen Evolution*. Cambridge University Press, New York, pp. 39–74.

Ruwende, C. *et al.* (1995). Natural selection of hemi- and heterozygotes for G6PD deficiency in Africa by resistance to severe malaria. *Nature*, **376,** 246–249.

Snow, R.W. *et al.* (2005). The global distribution of clinical episodes of *Plasmodium falciparum* malaria. *Nature*, **434,** 214–217.

Tishkoff, S.A. and Verrelli, B.C. (2004). G6PD deficiency and malarial resistance in humans: insights from evolutionary genetic analyses. In: K.R. Dronamraju (ed), *Infectious Disease and Host-Pathogen Evolution*. Cambridge University Press, New York.

Volkman, S.K., *et al.* (2001). Recent origin of *Plasmodium falciparum* from a single progenitor. *Science*, **293,** 482–484.

Vuilliamy, T., Mason, P., and Luzzatto, L. (1992). The molecular basis of glucose-6-phosphate dehydrogenase deficiency *Trends Genet.*, **8,** 138–143.

Weatherall, D.J. (2004). J.B.S. Haldane and the malaria hypothesis. In: *Infectious Disease and Host-Pathogen Evolution* (K.R. Dronamraju, ed). Cambridge University Press, New York, pp. 18–36.

Weatherall, D.J. and Clegg, J.B. (eds) (2001). The Thalassemia Syndromes (4th ed). Blackwell Science, Oxford.

Weatherall, D.J. and Clegg, J.B. (2002). Genetic variability in response to infection. In *Malaria and after. Paeds. Immunity,* **3,** 331–337.

Williams, T.N. *et al.* (1996). High incidence of malaria in-thalassemic children. *Nature*, **383,** 522–525.

Williams, T.N. *et al.* (2002). The membrane characteristics of *Plasmodium falciparum*-infected and -uninfected heterozygous thalassemic erythrocytes. *Br. J. Hematology*, **118,** 663–670.

J.B.S. Haldane (1892–1964)

Krishna R. Dronamraju

In view of the leadership of J.B.S. Haldane, the chief architect who formulated the "malaria hypothesis," it is of interest to review his contributions to genetics briefly, especially to the aspects of evolution and human genetics. He was a remarkable scientist. He possessed no formal academic qualification in science, yet he became one of the twentieth century's most influential scientists. He was a polymath whose intellectual versatility covered many disciplines, including physiology, genetics, biochemistry, biometry, statistics, cosmology, and philosophy, to name a few. Furthermore, he was a most skilled and prolific popularizer of science whose articles in the popular press covered even more disciplines. The following account is a brief summary of Haldane's contributions to genetics. His important work in physiology, biochemistry, biometry, and other fields is not included here.

He was the author of over 400 research papers (for most of which he was the sole author), 24 books, numerous popular articles, book-reviews, and political speeches. His intellectual productivity was even more remarkable when we consider that the computer and the word processor did not even exist during his lifetime.

Haldane not only understood several scientific disciplines, but he also made fundamental contributions to many of them. In many fields, his ideas, concepts, and methods shaped the course of a given field, often transforming it profoundly and introducing a rigor that did not exist before. He frequently made connections between disparate fields of science, which others (so-called professional scientists in those disciplines) had missed, thereby introducing new points of view and opening whole new fields of science. Examples include the connection between malaria and thalassemia (Haldane 1949a,b), and the connection between anaerobic environment and the origin of primitive life (Dronamraju, 1968).

Krishna R. Dronamraju • Foundation for Genetic Research, P.O. Box 27701-0, Houston, Texas 77227.

Malaria: Genetic and Evolutionary Aspects, edited by Krishna R. Dronamraju and Paolo Arese, Springer, New York, 2006.

John Burdon Sanderson Haldane (1892–1964), or "JBS" as he was widely known was the only son of an Oxford University physiologist, John Scott Haldane (1860–1936), who was distinguished for his contributions to respiratory physiology. His distinguished ancestors included his uncle Lord Haldane who was Chancellor of the Exchequer and his great uncle, Burdon Sanderson, who was the first Waynflete Professor of Physiology at Oxford University.

The younger Haldane was groomed by his father from his childhood to follow the scientific tradition. J.B.S.'s intellectual precocity was legendary. At the age of 3, when he was wounded in a fall and was bleeding, he was reported to have asked the doctor whether his blood contained oxyhemoglobin or carboxyhemoglobin. Before he was 10, he used to accompany his father into coal mines to collect air samples and on one occasion, readily calculated a set of log tables when asked to do so by his father who had forgotten to bring his own. He was educated at Eton and Oxford University, graduating in Classics in 1915. But, while he was a student at Oxford he had already published two physiological papers in collaboration with his father and also an important paper on one of the first cases of genetic linkage in the vertebrates (Haldane *et al.*, 1915). He possessed a brilliant mind and prodigious memory and was fond of quoting, in his later years, extensive passages from Greco-Roman classics, Hindu epics (e.g., *Ramayana* and *Mahabharata*), and excertps from psalms and various other religious texts, during international scientific conferences.

During his long career, Haldane was successively identified as a physiologist, biochemist, geneticist, biometrician, and a statistician, among others. He was a Fellow in physiology at New College, Oxford during the years 1919–1922. He was appointed as the Sir William Dunn Reader in the Biochemistry Department at Cambridge University by Sir Gowland Hopkins in 1923 and served with distinction for 10 years when he resigned, in 1933, to accept the Chair of Genetics (1933–1937) and later Biometry (1937–1957) at University College, London. In July 1957, he moved to India when offered a Research Professorship at the Indian Statistical Institute in Calcutta. Haldane directed research in India in several disciplines but especially in human genetics. He continued his mathematical investigations in population genetics. This period of Haldane's life was discussed in detail in my book: *Haldane: The Life and Work of J.B.S. Haldane with special reference to India* (Dronamraju 1985).

Haldane contributed original ideas, concepts, and methods to almost all aspects of genetics. The following account touches upon only a few highlights of his many important contributions to genetics. A fuller account can be found in various biographical accounts (Dronamraju, 1985, 1987, 1990, 1992a,b, 1993, 1995, 1996, 2004).

Photo of J.B.S. Haldane taken in Madison, Wisconsin, shortly before his death in 1964. (Photo was taken by Dr. Klaus Patau.)

1. Population Genetics

Haldane served in the Black Watch during World War I, and soon after returning to civilian life, published an important paper on the estimation of linkage between loci based on recombination, later known as "mapping functions." In the same paper, he proposed the term "centimorgan" or *cM*, which is still used today (Haldane, 1919). But his major interest during the 1920s was the development of an extensive mathematical theory of natural selection. In a series of mathematical papers on a quantitative theory of natural selection, Haldane derived equations to examine the consequences of selection and mutation under several hypothetical genetic circumstamces, such as inbreeding, outcrossing, dominance, recessivity, partial penetrance, epistasis, and linkage, etc. For instance, he showed that for a dominant gene with a selective advantage of 0.001, a total of 6920 generations is required to change the gene frequency from 0.001 to 1%, a total of 4819 generations is required to change from 1 to 50%, a total of 11,664 generations is required to change from 50 to 99%, and a total of 3,09,780 generations is required to change from 99 to 99.999%. He also examined the consequences of selection for quantitative traits and very rare characters. Haldane's investigations during that period (1924–32) were summed up in his book *The Causes of Evolution* (Haldane, 1932). His work laid the foundation of what later came to be called *population genetics* to which Fisher (1930) and Wright (1931) also contributed independently. However, unlike Fisher and Wright, Haldane contributed to the foundations of human genetics quite extensively. His knowledge of demographic genetics, biochemical genetics, and several other branches of genetics was far greater than that of his contemporaries.

Haldane's contribution to population genetics differed from those of Fisher and Wright in several respects. He considered both the statics and the dynamics of evolution, and much of his early work on gene frequency changes was deterministic. In one of his papers on the mathematical theory of natural selection, Haldane (1927a) dealt with a stochastic problem, investigating the probability of fixation of mutant genes. Haldane showed that the probability that a single mutation (with selective advantage k) will ultimately become established is only about $2k$ if dominant and only about $\sqrt{k/N}$ if completely recessive, where N is the population size. Although this analysis was greatly extended by the later works of Fisher, Wright, and Kimura, Haldane was the first to tackle this aspect of population genetics.

In another important investigation, Haldane (1931) anticipated Wright's shifting balance theory. His paper on "metastable populations" contained an elegant demonstration that mutant genes, which are harmful singly, may become advantageous in combination. He then showed that, for *m* genes, a population can be represented by a point in *m*-dimensional space. He argued that the process of speciation can result from a rupture of the metastable equilibrium and that such ruptures may be more likely to occur in small isolated communities. During his later years in India, he returned to many of these problems in population genetics, some in collaboration with Jayakar, investigating the conditions for polymorphism under selection of varying direction (Haldane and Jayakar 1963), the elimination of double dominants in large random mating populations, polymorphism due to selection depending on the composition of a population, and solutions to some problems in population genetics that were first considered by Haldane long time ago. One of his last papers with Jayakar, gave in a very elegant form the conditions for stability of an intermediate gene frequency at a sex-linked locus. These are summarized by Dronamraju (1985).

2. Beanbag Genetics

Towards the end of his life, Haldane (1964) wrote a remarkable essay, entitled "A defense of beanbag genetics," justifying the value of theoretical population genetics, which was founded by him, R.A. Fisher, and Sewall Wright. Evolutionary biologist, Ernst Mayr (1963), had earlier questioned the importance of theoretical population genetics and its relevance to evolutionary biology. Haldane (1964) responded by saying that a mathematical theory may be regarded as a kind of scaffolding within which a reasonably secure theory expressible in words may be built up. He stated further that only algebraic arguments can be decisive in some situations and adequate field data were not forthcoming to test the theoretical models. Haldane wrote that he made certain simplifying assumptions, which enabled the framing of biological questions in a form that would suit mathematical analyses. He added that the mathematical methods employed by himself, Fisher, and Wright would not impress professional mathematicians as they were simple by mathematical standards.

3. Terminology

There are at least three instances of Haldane's contribution to the terminology of genetics and evolutionary biology. He introduced the term *centimorgan*, or *cM*, as a unit of chromosome map distance, deriving a relationship between the distance (*mapping function*) between two loci and their crossover value (Haldane, 1919). A second instance was his introduction of the terms *cis* and *trans* into genetics from biochemistry, replacing the terms first introduced by his mentor William Bateson—*coupling* and *repulsion*, which were in vogue until then (Haldane, 1941a). Haldane's (1949a) contribution to the terminology of evolutionary biology included the term *Darwin* as a measure of evolutionary rate on the basis of changes in tooth size in fossil horses. For example, it can be an increase or decrease of size by a factor of e per million years, or an increase or decrease of 1/1000 per 1000 years. The horse rates would range around 40 millidarwins. Haldane wrote that the unit for the character may be a unit increase in the natural logarithm of a variate, or alternatively one standard deviation of the character in a population at a given horizon. Haldane's paper was written at the invitation of Ernst Mayr, who, as the first editor of Evolution, was then soliciting papers from eminent evolutionary biologists.

4. Human Genetics

During the period, 1930–1964, Haldane published numerous papers on all aspects of the genetics of man, laying the foundation for what later came to be called "human or medical genetics." His methods were mathematical and statistical. Haldane's analysis and insight moved the field of human and medical genetics forward in the era of premolecular-biology. The subjects covered by Haldane include: the formal analysis of human pedigrees, especially his maximum likelihood method of estimating the true proportions of affected offspring in families with recessive hereditary diseases (Haldane 1932a), the first human mutation rate (for hemophilia) (Haldane 1932b, 1935), the first human gene map for hemophilia and color blindness on the X chromosome (Haldane and Bell 1937, Haldane and Smith, 1947), relation of modifying genes to age-at-onset variation (Haldane, 1941b), analysis of heredity–environment interaction (Haldane 1946), the role of infectious disease in evolution (Haldane, 1949b,c), and the measurement of natural selection (1954). Other papers of interest include a study of the impact of inbreeding on the spread of sex-linked genes in human populations (Haldane and Moshinsky, 1939) and the dysgenic effect of induced recessive mutations (Haldane, 1947). In his later years in India, he took special interest in the genetics of human populations of that region (Dronamraju and Haldane 1962, Haldane and Jayakar 1962).

"In 1954, Haldane proposed a measurement of the intensity of selection, $I = \ln s - \ln S$, where s is the fitness of the optimum phenotype and S is that of the whole population." By applying Karn and Penrose's (1951) data on human birth–weight distribution, Haldane found that 58% of all deaths were selective on the sole criterion of weight, and the intensity was $I = 0.0240 \pm 0.004$. The effect of this natural selection on population was to increase the mean birth weight from 7.06 to 7.13 pounds, or about 1% but to decrease the standard deviation of the birth weights by 10%. The effect of selection in decreasing the variance was far greater than its effect in increasing the mean.

Among Haldane's ideas that had a significant impact on the growth of genetics, were his early emphasis that the biochemical interpretation of gene action is more fundamental than was the practice until then. His paper on the genetic basis of human chemical individuality must be regarded as a milestone in human–biochemical genetics and as a connecting link between the early work of Archibald Garrod (1909) on human–biochemical disorders and the later work of Harris (1953) and others on human biochemical genetics (Haldane, 1937b). He was among the early geneticists who enunciated the gene–enzyme concept, "The chemist may regard them (genes) as large nucleoprotein molecules, but the biologist will perhaps remind him that they exhibit one of the most fundamental characteristics of a living organism: they reproduce themselves without any perceptible change in various different environments ... in some cases we have very strong evidence that they produce definite quantities of enzymes, and the members of a series of multiple allelomorphs produce the same enzyme in different quantities" (Haldane, 1920).

5. Genetic Load Theory

The general idea of genetic loads is based on a paper by Haldane (1937), "The effect of variation on fitness." He showed that the effect of mutation on population fitness depends mainly on the mutation rate and not on the deleteriousness of the individual mutants. Several years later, Muller (1950) discussed the problem under the title: "Our load of mutations". The theory of genetic loads was chiefly developed by Crow (1958) and Crow and Kimura (1965). Haldane's (1937a) paper gave the first basis for assessing the impact of mutation on the population. It also showed that any increase in mutation rate would have an effect on fitness ultimately equal to this increase. This principle provided a basis for various assessments of the genetic effect of radiation at a time when the question first became one of social and political importance.

6. Immunogenetics

An important contribution of Haldane (1933) was the clarity he introduced into transplantation and immunological genetics. He suggested for the first time that the transplantation factors for tumors identified by C.C. Little and George Snell at the Jackson Laboratory in Bar Harbor were simply antigens controlled by genes that are present and active in both tumors and normal tissues. Haldane's contributions to immunogenetics were summarized by Mitchison (1968). The maintenance of heterozygosis for genes controlling antigens was of great interest to Haldane. He examined the expected rate of approach to homozygosis in inbreeding (Haldane 1936, 1956) and during backcrossing that was specially designed to produce co isogenic strains of mice (Haldane and Snell 1948). He foresaw correctly that inbreeding will not produce complete genetic uniformity as was demonstrated by later work on skin grafts among inbred mice.

Among other ideas was Haldane's (1944) suggestion that the Rhesus polymorphism lacked any satisfactory explanation and was therefore of recent origin. His analysis of maternal–fetal incompatibility showed that the rhesus factor could be in unstable equilibrium when the gene frequency was 1/2, however, most populations would tend to become either Rh-positive or Rh-negative (Haldane 1942). Following the discovery of the genetic basis of human fetal erythroblastosis by Landsteiner and Wiener, Haldane pointed out that the modern populations of Europe were the result of crossing between populations who possessed a majority of Rh-positive genes and populations who had a majority of Rh-negative genes. Furthermore, two populations—one in Switzerland and another in Spain—were discovered to have a majority of Rh-negative genes.

Haldane suggested that the Rh + genes come from Asia and Rh- from Europe. Thus finding thats the Basques, which are thought to be the closest descendants of the original Europeans, and high frequency Rh- alleles supports the idea.

Haldane suggested that a high rate of mortality in their offspring would indicate that such differences would constitute a barrier to crossing, and play some part in preventing hybridization between mammalian species (Haldane 1964). On the other hand, he suggested that such common diseases as amaurotic idiocy, microcythemia, and sickle cell (which are too frequent to be attributed to mutation) might be due to selection in favor of heterozygotes leading to stable equilibria (Haldane 1949c).

Haldane's impact on immunogenetics was evident through his colleagues and pupils (see Haldane 1956, Haldane and Snell 1948). One of them was his nephew N.A. Mitchison who described Haldane's impact on his work in a contribution to the memorial tribute, which I have edited shortly after Haldane's death (Mitchison 1968). Many years ago, in 1933, Haldane recruited Peter Gorer to join his Department at University College,

London. That move by Haldane proved to be a very important step in advancing immunogenetics. Gorer at first discovered three blood groups among the strains of mice. One of these was named "antigen II", which later became the H-2 locus, that landmark of transplantation biology.

7. Sociobiology

Of particular interest was Haldane's analysis of socially valuable but individually disadvantageous characters (e.g., as in social insects, such as honey bees and ants). Haldane's analysis of altruism led to the development of sociobiology many years later. Haldane (1932a) stated that socially valuable but individually disadvantageous characters can only spread through the population if the genes determining them are carried by a group of related individuals. The chances of these individuals for leaving offspring are increased by the presence of these genes in an individual member of the group whose individual viability is lowered by these very genes. Haldane (1932a) gave two examples. (a) With respect to the inherited trait *broodiness* in poultry, Haldane stated that a broody hen is more likely to have a shorter life than a nonbroody hen because she is more likely caught by a predator while sitting. However, the nonbroody hen will not rear a family, so genes determining this character will be eliminated in nature. Selection may strike a balance between these two extremes—neither a too devoted parent nor one that abandons her young at the slightest danger would be represented in posterity. On the other hand, one that does so only under intense stimulus will live to rear another family. (b) The other example concerns the devotion and self-sacrifice among social insects that may be of biological advantage. In a beehive, the same set of genotypes is represented in the workers and young queens. Any behavior pattern in the workers ("however suicidal it may be") that is of advantage to the hive, will promote the survival of the queens, and thus tend to be favorably selected. On the other hand, genes causing unduly altruistic behavior in the queens would tend to be eliminated. Haldane concluded that the biological advantages of altruistic conduct only outweigh the disadvantages if a substantial proportion of the tribe behaves altruistically. Many years later, Hamilton's (1964) analysis of sociobiology was based on Haldane's ideas.

8. Daedalus and Eugenics

Haldane was a highly skilled popularizer of science and its social applications. Through his numerous articles in the popular magazines and newspapers he exercised great influence in educating millions of people. But none matched his first book—a slim volume entitled *Daedalus, Or*

Science and the Future, which was first published in London in 1923 and in New York a year later. It was an expanded version of his address to Heretics at Cambridge University which was delivered on 4th Feb 1923. The biological predictions in Haldane's *Daedalus* were in the form of hypothetical essay on the influence of biology on history during the 20th century which will (" it is hoped") be read by a rather stupid undergraduate member of this university to his supervisor during his first term 150 years hence. *Daedalus* was a remarkable statement, which was a prediction of what might happen in science in the far distant future and at the same time an appeal to encourage certain lines of biological research, especially genetic engineering, as well as a warning against possible misuse or excessive zeal in applying science to solve social problems. It was a bold document, which proclaimed what scientific revolution might bring forth in the most private aspects of life, death, sex, and marriage, in an era when even the mention of 'birth control' in public media caused an uproar (see Huxley 1970, p. 151; and Dronamraju 1993a, p. 122). Haldane predicted the widespread practice of eugenic selection, *in vitro* fertilization, manipulation of the human genome, and routine production of offspring with exceptional qualities in music, sports, and virtue. He prophesied the widespread use of psychotropic drugs and numerous other biological and therapeutic interventions to modify the psychological behavioral, as well as the biochemical condition of the human body and mind. *Daedalus* became a sensational bestseller overnight and the American edition was reprinted five times in 1924. Many years later, in 1932, Aldous Huxley's *Brave New World* was published. It was simply a fictionalized version of Haldane's *Daedalus* but without any acknowledgement to the source. In response to *Daedalus*, a contrary point of view was expressed by Bertrand Russell (1924) who argued that science is no substitute for virtue.

Daedalus firmly established Haldane as an outstanding popular science writer (also see Haldane 1993b). Throughout his life, he wrote numerous popular scientific articles, which appeared in the newspapers and magazines in several countries, and were later published in collected volumes, such as *Possible Worlds and Other Essays* (1927b), *The Inequality of Man and Other Essays* (1932c), *Science in Everyday Life* (1940a), *Keeping Cool and Other Essays* (1940b), *Science Advances* (1947), and several other books.

References

Crow, J.F. (1958). The possibilities for measuring selection intensities in man. *Hum. Biol.,* **30,** 1–13.

Crow, J.F. and Kimura, M. (1965). The theory of genetic loads. In: Geerts, S.J. (ed) *Genetics Today, Proc. XI Intl. Cong. Genet.* vol. 3, The Hague, pp. 495–505.

Dronamraju, K.R. (ed) (1968). *Haldane and Modern Biology*. Johns Hopkins University Press, Baltimore.

Dronamraju, K.R. (1985). *Haldane: The Life and Work of JBS Haldane with special reference to Haldane*. Aberdeen University Press, Aberdeen. U.K.

Dronamraju, K.R. (1987). On some aspects of the life and work of J.B.S. Haldane in India. *Notes Records Roy. Soc.* **41,** 211–223.

Dronamraju, K.R. (ed) (1990). *Selected Genetic Papers of JBS Haldane*. Garland Publishing Co., New York.

Dronamraju, K.R. (1992a) Historical essay: J.B.S. Haldane (1892–1964): Centennial Appreciation of a Polymath. *Am. J. Hum. Genet*, **51,** 885–889.

Dronamraju, K.R. (1992b) Profiles in Genetics. J.B.S. Haldane (1892–1964): A Centennial Appreciation. *J. Hered.*, **83,** 163–168.

Dronamraju K. R. (1993a) If I am To Be Remembered: The Life and work of Julian Huxley with selected correspondence. World Scientific, London.

Dronamraju, K.R. (1993b). J.B.S. Haldane's (1892–1964) biological speculations. *Hum. Gene Ther.* **4,** 303–306.

Dronamraju, K.R. (ed) (1995). *Haldane's Daedalus Revisited*. Oxford University Press, Oxford, U.K.

Dronamraju, K.R. (1996). Haldane's dilemma and its relevance today. *Curr. Sci.*, **70**, 1059–1061.

Dronamraju, K.R. (ed) (2004). *Infectious Disease and Host-Pathogen Evolution*. Cambridge University Press, New York.

Dronamraju, K.R. and Haldane, J.B.S. (1962). Inheritance of hairy pinnae. *Am. J. Hum. Genet.*, **14,** 102–103.

Fisher, R.A. (1930). *The Genetical Theory of Natural Selection*. Clarendon Press, Oxford.

Garrod, A.E. (1909) Inborn Errors of Metabolism. London: Oxford University Press.

Haldane, J.B.S., Sprunt, A.D. and Haldane, N.M. (1915) Reduplication in mice. *J. Genet.*, **5,** 133–135.

Haldane, J.B.S. (1919). The combination of linkage values, and the calculation of distances between the loci of linked factors. *J. Genet.*, **8,** 299–309.

Haldane, J.B.S., Sprunt, A.D., and Haldane, N.M. (1915) Reduplication in mice. *J.Genet.*, 5: 133–135.

Haldane, J.B.S (1920) Some reason to work on heredity. Trans. Oxford Univ. Jr. Sci. Club, 1: (3) 3–11.

Haldane, J.B.S. (1923). Daedalus, Or Science and the Future. Kegan Paul, London.

Haldane, J.B.S. (1924). A mathematical theory of natural and artificial selection. Pt. 1, *Trans. Camb. Phil. Soc.*, **23,** 19–41.

Haldane, J.B.S. (1927a). A mathematical theory of natural and artificial selection. Pt. V. Selection and mutation. *Proc. Camb. Phil. Soc.*, **23,** 833–844.

Haldane, J.B.S. (1927b). *Possible Worlds and Other Essays*. Chatto & Windus, London.

Haldane, J.B.S. (1931). A mathematical theory of natural and artificial selection. Part VIII. Metastable populations. *Proc. Camb. Phil. Soc.*, **27,** 137–142.

Haldane, J.B.S (1932a) A method for investigating recessive characters in man. *25:* 251– 256.

Haldane, J.B.S. (1932b). *The Causes of Evolution*. Longmans Green, London.

Haldane, J.B.S. (1932c). *The Inequality of Man and Other Essays*. Chatto & Windus, London.

Haldane, J.B.S. (1933). The genetics of cancer. *Nature*, **132,** 265–267.

Haldane, J.B.S. (1935). The rate of spontaneous mutation of a human gene. *J. Genet.*, **31,** 317–326.

Haldane, J.B.S. (1936). The amount of heterozygosis to be expected in an approximately pure line. *J. Genet.*, **32,** 375–391.

Haldane, J.B.S. (1937a). The effect of variation on fitness. *Am. Nat.*, **71,** 337–349.

Haldane, J.B.S., (1937b) Biochemistry of the individual. In: In Needham, J. and Greene D.E. (Eds.) Perspectives in biochemistry Cambridge: Cambridge University press.

Haldane, J.B.S. (1940a). *Science in Everyday Life*. Macmillan, New York.

Haldane, J.B.S. (1940b). *Keeping Cool and Other Essays*. Chatto & Windus, London.

Haldane, J.B.S. (1941a). *New Paths in Genetics*. Allen and Unwin, London.

Haldane, J.B.S. (1941b). The relative importance of principal and modifying genes in determining some human diseases. *J. Genet.*, **41,** 149–157.

Haldane, J.B.S. (1942). Selection against heterozygosis in man. *Ann. Eugen.*, **11**, 333–343.
Haldane, J.B.S. (1944). Mutation and the Rhesus reaction. *Nature*, **153,** 106–108.
Haldane, J.B.S., (1946) The interaction of nature and nurture. Ann. Eugen., 13, 197–202
Haldane, J.B.S. (1947). *Science Advances*. Allen & Unwin,, London.
Haldane, J.B.S. (1949a). Suggestions as to quantitative measurement of rates of evolution. *Evolution*, **3,** 51–56.
Haldane, J.B.S. (1949b). The rate of mutation of human genes. In: *Proc. VIII Intl. Cong. Genetics, Hereditas*, **35,** 267–273.
Haldane, J.B.S. (1949c). Disease and evolution. *La Ricerca Scientifica*, **19,** 2–11.
Haldane, J.B.S. (1954). The measurement of natural selection. *Caryologia*, **6**(Suppl.), 480–487.
Haldane, J.B.S. (1956). The detection of antigens with an abnormal genetic determination. *J. Genet*., **54**, 54–55.
Haldane, J.B.S. (1964). A defense of beanbag genetics. *Persp. Biol. Med.* **7,** 343–359.
Haldane, J.B.S. and Bell, J. (1937). The linkage between the genes for colour-blindness and haemophilia in man. *Proc. Roy. Soc., Lond. B* **123,** 119–150.
Haldane, J.B.S. and Moshinsky, P. (1939). Inbreeding in Mendelian populations, with special reference to human cousin marriage. *Ann. Eugen.*, **9,** 321–340.
Haldane, J.B.S. and Smith, C.A.B. (1947). A new estimate of the linkage between the genes for colour-blindness and haemophilia in man. *Ann. Eugen.*, **14,** 10–31.
Haldane, J.B.S. and Snell, G.D. (1948). Methods for histocompatibility of genes. *J. Genet.*, **49,** 104–108.
Haldane, J.B.S. and Jayakar, S.D. (1962). An enumeration of some human relationships. *J. Genet.*, **58,** 81–107.
Haldane, J.B.S. and Jayakar, S.D. (1963). Polymorphism due to selection of varying direction. *J. Genet.*, **58**, 237–242.
Hamilton, W.D. (1964). The evolution of social behavior. *J. Theor. Behav*., **7,** 1–16; **7,** 17–52.
Harris, H. (1953). An introduction to Human Biochemical Genetics. London: Cambridge University Press.
Huxley, Aldous (1932). *Brave New World*. Penguin, New York.
Huxley, Julian (1970) Memories. Harper & Row, New York
Karn, M.N. and Penrose, L.S. (1951). Birth weight and gestation time in relation to maternal age, parity, and infant survival. *Ann. Eugen*., **16,** 147–164.
Mayr, E. (1963) *Animal Species and Evolution*. Cambridge, Mass.: Harvard University Press.
Mitchison, N.A. (1968). Antigens. In: Dronamraju, K.R. (ed) *Haldane and Modern Biology*. Johns Hopkins University Press, Baltimore.
Muller, H.J. (1950) Our load of mutations.2, 111-140
Russell, B. (1924). *Icarus, or the Future of Science*. E.P. Dutton, New York.
Wright, S. (1931). Evolution in Mendelian populations. *Genetics*, **16,** 97–159.

Removal of Early Parasite Forms from Circulation as a Mechanism of Resistance Against Malaria in Widespread Red Blood Cell Mutations

Paolo Arese, Kodjo Ayi, Aleksei Skorokhod and Franco Turrini

1. Introduction

Anecdotal evidence has long indicated that native populations living in malarial areas were resistant against the disease, while nonadapted foreigners were usually highly susceptible, unless proper treatment was available. For example, according to some historians outbreaks of malaria epidemics have been considered responsible for the decline of the Roman Empire, for the unsuccessful outcome of the Crusades and, on a minor scale, for the Gallipoli disaster in World War I (Spielman and D'Antonio, 2001; Sallares, 2001). High susceptibility of foreigners to malaria was also utilized by the ruling class or the judiciary as a way of punishment and a kind of cruel and inescapable death penalty for unwanted individuals or criminals. For example, the Romans used to send political opponents to Sardinia and the rulers of Siena used to ban criminals to the Maremma, a coastal area in Tuscany, both heavily malarial areas until modern times.

Evidence of widespread resistance in "native" populations struck by parasitic or bacterial diseases with endemic character has been formalized by Haldane and other geneticists in 1930s, making use of the

---

Paolo Arese, Kodjo Ayi, Aleksei Skorokhod and Franco Turrini • Department of Genetics, Biology and Biochemistry, University of Torino Medical School, Torino, Italy.

Malaria: Genetic and Evolutionary Aspects, edited by Krishna R. Dronamraju and Paolo Arese, Springer, New York, 2006.

Darwinian categories of natural selection by fitness. Natural selection, the most significant evolutionary agent, causes differential survival of some genotypes at the loss of others. Haldane introduced the concept of "balanced polymorphism" and "selective advantage of the heterozygote." He pointed out that infectious diseases influence the frequency of genes that provide resistance (Haldane, 1949a,b). Expansion of certain genes is either due to increase in reproductive rate or decrease in mortality, or a combination of both. Considering genes that lower mortality, they enhance evolutionary survival and would tend to increase in a population and replace their "normal" alleles. Heterozygote advantage and balanced polymorphism means that a mutant allele provides a selective advantage in the heterozygote but has lethal or severe outcome in the homozygote. In other words, the hematologic disadvantage of the homozygote is balanced by the heterozygote advantage of protection from malaria. In this case, the frequency of heterozygotes increases until it is balanced by the negative effects of the homozygous state.

Proof of the evolutionary advantage of putative malaria protective mutations in the heterozygous state requires field studies showing that the distribution of any protective condition is geographically coincident with the distribution of the disease, and the frequency of the protective condition is correlated with the severity and frequency of the disease. A second requirement is case-control studies, showing that affected heterozygotes have lower mortality or less severe symptoms compared to nonmutant subjects. A third requirement is that a mechanism is found to explain why the condition brings about less severe symptoms and lower mortality.

The increased occurrence of specific hereditary conditions in "native," adapted populations resistant against malaria became possible after the identification and description of diseases affecting the human red blood cell(s) (RBC, RBCs), the demonstrations of their hereditary character, and the evidence for a distinct geographic coincidence between the past or present distribution of malaria and one or several mutations. Latter proof depended on the introduction of cheap, reliable, and easy-to-perform tests that allowed detection of those defects in heterozygotes even in absence of patent hematological alterations and other pathologic manifestations. Availability of such mass-testing methods allowed the screening of large numbers of individuals in field studies performed even in secluded regions without hospitals or health care facilities. Pioneering studies on the distribution and frequency of the thalassemias, the first RBC disease shown to be connected with malaria were performed by the Italian hematologists Enzo Silvestroni and Ida Bianco during and shortly after World War II (Bianco-Silvestroni, 2002). The second hereditary RBC disorder found to be connected with malaria resistance was sickle-cell disease, screened in Africa during the late 1940s (Beet, 1947; Allison, 1954, 1956, 1964). The next RBC disorders now generally considered as malaria-defensive (G6PD-deficiency, hemoglobin E, hemoglobin C, and ovalocytosis) were described and mass-screened in the next few years (Beet, 1947; Roberts *et al.*, 2004; Greene, 1993; Chotivanich *et al.*, 2002; Agarwal *et al.*, 2000; Modiano *et al.*, 2001; Serjeantson *et al.*, 1977).

In the following, we will briefly summarize geographical and epidemiological evidence indicating that malaria was most likely the shaping factor for the peculiar distribution and frequency of those RBC diseases. We will then show the protective efficacy of those conditions and will critically discuss the current way to mechanistically explain protection. We will then propose enhanced phagocytic removal of early parasite forms from the circulation as an alternative explanation for resistance in subjects with major RBC mutations. We will describe the modifications in the RBC membrane elicited/enhanced by the developing parasite and show that those parasite-induced changes modify the surface of the parasitized RBC in much the same way as normal senescence or oxidative insults modify nonparasitized RBC and induce their phagocytosis. We will show data obtained *in vitro* showing enhanced phagocytosis of rings developing in mutant RBCs and discuss direct and indirect evidence, indicating that enhanced phagocytosis of those rings is indeed occurring *in vivo*, and we will finally discuss why phagocytosis of early parasite forms may be advantageous to the host and provide a generally valid model of protection.

2. Malaria is Responsible for High Frequency and Regional Distribution of Major Protective RBC Mutations: Geographical Evidence

The first large-scale screening campaigns of RBC disease that later turned out to be protective against malaria were conducted by the Italian doctor couple Enzo Silvestroni and Ida Bianco. The first surveys were done before, during, and shortly after World War II in several Italian regions (Bianco-Silvestroni, 2002). Silvestroni and Bianco utilized the simple Simmel technique, based on the detection of increased osmotic resistance of certain blood samples by visually testing hemolysis using a series of sodium chloride solutions with different osmolarity. Normal RBCs completely lyse at a sodium chloride concentration of 0.36%, while "microcytemic" RBCs stay intact. The term "microcytemia" was adopted by Silvestroni and Bianco because RBCs with increased osmotic resistance had invariably a –30% lower volume than normal RBC. The screening activity concentrated in the regions with high numbers of patients with Cooley anemia, since Silvestroni and Bianco rightly assumed "microcytemia" to be a milder form of Cooley's disease. The frequency of microcytemia was very high in Sardinia and in the Po river delta region, and high in specific coastal regions of Southern Italy and Sicily. Sardinia was a heavily malaric region until the complete malaria eradication after the 1948–1953 campaigns funded and coordinated by the Rockefeller Foundation. In Sardinia, data from very accurate malariological records and microcytemia screenings were combined on a microgeographic scale. The first to perform screenings in lowland and mountainous villages was the Italian geneticist R. Ceppellini. In now forgotten studies he and his co-workers described a high prevalence of microcytemia in the Sardinian coastal areas and a distinct cline with lower frequencies in mountainous villages, where malariological records indicated higher and significantly lower incidence of malaria, respectively (Ceppellini, 1955; Carcassi *et al.*, 1957). These early studies were later expanded and confirmed by Siniscalco *et al.* (1961, 1966). Microgeographical studies were later conducted in Papua New Guinea where researchers from the Oxford group found similar gradients correlating high gene frequencies for alpha-thalassemia in low-altitude areas known historically to have been malarious, and low frequencies in nonmalarious areas (Flint *et al.*, 1986, 1993; Hill, 2001; Yenchitsomanus *et al.*, 1986). Further studies and thorough analysis of the gene frequency of alpha-thalassemia led to at least two important observations: the gene frequency was low in nonmalarious areas and proportionally correlated to the incidence of malaria, with north–south and high–low-altitude clines paralleling the cline in the historical incidence of malaria; second, the frequencies of neutral genetic markers for which there is no evidence for protection did not show a similar correlation (Flint *et al.*, 1986, 1993; Hill, 2001; Yenchitsomanus *et al.*, 1986).

On a macrogeographical scale, there is overlapping between malaria occurrence and high-frequency of Hb AS occurrence. The Hb S gene is frequent in Sub-Saharan Africa and in India, where Lehmann first described the high frequency of Hb AS in tribal groups in Southern India in 1952 (Lehmann and Cutbush, 1952). The incidence of Hb S gene in India has been later examined a series of studies, as reported in a recent review, showing high frequencies in many tribal groups throughout the subcontinent, particularly in Central and Western India (Roberts *et al.*, 2005). The association between geographic distribution and frequency of Hb AS and protection from severe malaria has been reported in few studies. For example, Ramasamy (Roberts *et al.*, 2005) noted that distribution of Hb AS was associated with a cline in the probable malaria occurrence from west to east. The occurrence of low–high altitude clines has not been examined, but the frequency of the Hb S gene in East Africa was correlated with the endemicity of malaria.

3. Epidemiological Evidence: Degree of Protection Afforded by RBC Mutations

In this section a brief account of frequent mutations with protective character is given, including the type of mutation and the degree of protection.

3.1. Hemoglobin AS (Sickle-Cell Trait)

The best studied protective mutation is hemoglobin AS (Hb AS). Hb S is due to a single point mutation in the gene for the beta chain with a valine-to-glutamate substitution at position 6. Homozygotes for Hb S are severely ill with sickle-cell anemia and die before reproductive age in absence of appropriate treatment, usually not available in Africa. The heterozygous condition (sickle-cell trait; Hb AS) was found to be very common in a number of epidemiological surveys performed in various parts of Sub-Saharan Africa, with gene frequencies exceeding 30% in many parts of the continent. Allison was the first to show that heterozygotes had less parasites in their blood and were less likely to die from malaria (Allison, 1954, 1956, 1964). These results were confirmed by a number of other studies (see Jones, 1997; Lell *et al.*, 1999; Livingstone, 1971). In every study performed in different African regions, mortality from cerebral malaria was distinctly lower in Hb AS individuals. Studies on the frequency of parasitemia were less conclusive but did show in general lower rates for Hb AS individuals. Summarizing older and more recent studies, several case-control studies have clearly shown that sickle-cell trait is 90% protective against cerebral malaria and severe anemia, the most frequent causes of death, and 60% protective against malaria forms leading to hospital admission.

3.2. Thalassemias

The thalassemias are disorders of hemoglobin production. Alpha-thalassemia affects the duplicated alpha-chain genes and causes underproduction of alpha-globin chains while beta-thalassemia affects the single beta-chain gene and causes abrogation or underproduction of beta-chains. Beta-thalassemia is caused by over 200 different mutations, ensuing in hematological disorders of varying gravity, from complete absence of beta-chain synthesis and death at infant age (beta-zero thalassemia), to mild or very mild hematologic consequences (beta-plus thalassemia). Due to gene duplication, the situation is more complex in alpha-thalassemia, where, e.g., homozygous state for alpha-plus thalassemia [–alpha/–alpha] produces a mild hypochromic anemia (Weatherall, 1998). The thalassemia syndromes are extraordinarily widespread all over the world, and in general their geographic distribution corresponds to that of present or previous malaria (Weatherall, 1997). A single case-control study in Africa has found a protection of 50% in patients hospitalized for malaria (Allen *et al.*, 1997). Protection was later confirmed for beta-thalassemia in Africa (Willcox *et al.*, 1983). Protection afforded by alpha-thalassemia has been the object of a series of important studies by the Oxford group (Allen *et al.*, 1997; Williams *et al.*, 2005) and by others (Mockenhaupt *et al.*, 2004). Case-control studies have shown 60% protection from severe malaria for alpha-plus thalassemia homozygotes (–alpha/–alpha) and 34% for alpha-plus thalassemia heterozygotes. Surprisingly, though, children with alpha-plus thalassemia were found to have increased incidence of mild uncomplicated *vivax* malaria and higher prevalence of splenomegaly, an index of malaria infection (Williams *et al.*, 1996).

3.3. Hemoglobin C

The gene for Hb C is allelic with that for Hb S but codes for lysine-to-glutamate substitution at position 6. The condition is much less severe than Hb S, and homozygous subjects are hematologically comparable with mild forms of thalassemia. Hb C is only found in certain populations of West Africa (Mali, Burkina Faso, and Northern Ghana) where the frequency of the allele is between 10 and 20%. Hb C has been object of recent studies that confirm older data and indicate clear-cut protection. Two case-control studies have indicated 80 and 93% protection against severe malaria in Hb C homozygotes, respectively (Agarwal *et al.*, 2000; Modiano *et al.*, 2001). Protection offered by heterozygotes was lower (29% reduction in risk of clinical malaria (Modiano *et al.*, 2001).

3.4. Hemoglobin E

Hb E is due to a single point mutation in the gene for the beta chain with a glutamate-to-lysine substitution at position 26. The gene occurs at high frequencies (e.g., in the range 19–25% in Thailand) in the eastern half of the Indian subcontinent and Southeast Asia, in geographical coincidence with extremely severe malaria burden. With an estimated occurrence in 30 million people (in 1980) it is one of the most prevalent hemoglobinopathies. The homozygous condition constantly accompanied by microcytosis, is hematologically irrelevant, although under circumstances, Hb E has increased susceptibility to precipitation and oxidant hemolysis, and reduced glutathione was lower in Hb E RBCs. Protection against *falciparum* malaria has not been substantiated clearly, while there is evidence that Hb E may protect against *Plasmodium vivax* infection. However, a recent study has shown that in patients hospitalized with malaria, patients with Hb E trait had fewer severe complications (Hutagalung *et al.*, 2000). Interestingly, it has been shown that malaria patients with Hb E treated with artemisinin (but not chloroquine) clear parasites from their blood more rapidly as compared to nonmutant controls (Hutagalung *et al.*, 2000), and parasitized Hb E RBCs were phagocytosed more intensely compared to parasitized normal control RBCs (Bunyaratvej *et al.*, 1986).

3.5. Glucose-6-phosphate dehydrogenase

Despite several reports showing no or not significant protection, the consensus is now in favor of a protective effect of G6PD deficiency in both heterozygotes and hemizygotes (Greene, 1993; Ruwende *et al.*, 1995). Studies in favor of protection have been performed in Africa (Allison and Clyde, 1961), indicating lower parasite densities in hemizygous males but not in heterozygous females, with parasite densities approximately equivalent to those observed in sickle-cell trait individuals. A second, frequently cited series of studies by Luzzatto *et al.* (1969) confirmed protection, however restricted to heterozygous females and not present in hemizygous males and homozygous females. In those studies, G6PD genotype had been determined in 709 children (aged 9 months to 6 years) admitted to hospital with acute febrile illness, and compared with 189 healthy young adults aged 14–20 years. It has been remarked that hemizygous and homozygous subjects may be less likely to become infected, but, in case of infection, are more likely to get a more severe malaria due to the greater vulnerability to oxidants of deficient RBCs. Probably, a study like that of Allison and Clyde (1961) performed on a whole unselected sample of children out of a whole community would not suffer from a selection bias possibly present in Luzzatto's studies. In a more recent study, the frequencies of the G6PD A-, G6PD A, and G6PD B alleles was measured in over 2,000 DNA samples collected in two large case-control studies of severe malaria in East and West Africa. This setup seems to offer a more sensitive measure of the effect of malaria protection alleles (Ruwende *et al.*, 1995). Results indicate about 46% protection against severe malaria and also significant protection against mild malaria in female heterozygotes. Male hemizygotes were associated with about 58% protection, supporting the notion that the degree of enzyme deficiency (more severe in hemizygotes than in heterozygotes) is central to the protective mechanism. The study by Ruwende *et al.* (1995) convincingly showed that the degree of protection by G6PD deficiency is lower compared to sickle-cell trait but equal or greater than that afforded by the thalassemias or some HLA variants.

3.6. Southeast Asian Ovalocytosis

Southeast Asian ovalocytosis (SAO) is caused by two linked mutations with a point mutation Lys56-to-Glu substitution and the deletion of a 9-aminoacid stretch affecting band 3, a major RBC membrane protein and small anion exchanger that also plays an important role in elimination of senescent or damaged RBCs and malaria-parasitized RBCs (see Section 5.1). Homozygosity for SAO is lethal during fetal development and heterozygotes are protected against severe malaria, as confirmed (Foo *et al.*, 1992) after initial conflicting data. Interestingly, protection is highly specific for cerebral malaria (Allen *et al.*, 1999).

4. A Critical Assessment of the Current Mechanism of Protection by Widespread RBC Mutations

The predominant explanation of resistance due to widespread RBC mutations found in the malaria literature (for example, see Friedman, 1978; Kaminsky *et al.*, 1986; Brokelman *et al.*, 1987; Senok *et al.*, 1997; Pattanapanyasat *et al.*, 1999) is that the mutant RBCs do not offer optimal conditions for invasion, or intracellular parasite development, RBC rupture, and merozoite release. We consider evidence provided by most of those studies as not fully satisfactory (Akide-Ndunge *et al.*, 2003). In general, *in vitro* studies are fraught with two common biases (Akide-Ndunge *et al.*, 2003). The first one resides in the impossibility to reproduce the very complex *in vivo* situation with the simplified experimental setup adopted *in vitro*, as shown by the comparisons in Table 3.1. The second bias stems from the generalized use of inadequate culture media. Comparison between RPMI1640, the medium of choice in most studies, and blood plasma *in vivo* shows that several amino acids are low in medium and may limit protein synthesis and parasite growth (Akide-Ndunge *et al.*, 2003). It is of note that *Plasmodium falciparum* utilizes only approximately 15% of amino acids derived from digestion of host hemoglobin (Krugliak *et al.*, 2002) and also relies on adequate supply from the medium. In addition, RPMI1640 is devoid of hypoxanthine and adenine, both essential for parasite growth and low in reduced glutathione. The limiting level of essential components of culture media may become critical because the vast majority of studies compared normal and mutant cells on a volume (hematocrit) basis and not on a cell number basis. Since both thalassemic and Hb AS RBCs are microcytic, comparisons were performed between samples containing significantly

Table 3.1. Summary of the Effect of Some Malaria-Protective Mutations on Invasion and Growth of *P. falciparum*

Study no.	Reference	Mutation	RBC status (time after withdrawal)	*P. falciparum* strain	Growth medium and culture details	Incubation conditions	Invasion (I) and Growth (G)	
1	Pasvol *et al.* (1978); Pasvol (1980)	Hb AS	Fresh patient blood	Not determined	75% Medium 199/25% AB serum; Ht in culture: approximately 3%	21% O_2, 5% CO_2 5% O_2, 5% CO_2 (low O_2)	No difference ↓ at low O_2	No difference ↓ at low O_2
2	Friedman (1978)	Hb AS	Used within 4 weeks	FMG/FCR-3	RPMI1640/25 mM Hepes/10% AB serum; daily medium change for 2 days, and then twice for 1–2 days; Ht in culture: 6%	Candle jar 18% O_2, 3% CO_2 3% O_2, 3% CO_2 (low O_2)	No difference until day 4 at 18% O_2 ↓↓ at low O_2	No difference until day 4 at 18% O_2 ↓↓ at low O_2
3	Roth *et al.* (1983)	Beta-thal trait	Used within 3–4 days	FCR-3	RPMI1640/25 mM Hepes/10% AB serum; Daily medium change; Ht in culture: 5%	Candle jar	No difference	No difference
4	Kaminsky *et al.*, 1986	Beta-thal trait	Not specified	FCB	RPMI1640/25 mM Hepes/10% serum; Ht in culture: not specified	Candle jar	No difference in I and G (days 1–5) No difference in I and G (days 1–10) if subcultivated, but↓↓ G (days 6–10) if not subcultivated	
5	Brockelman *et al.* (1987)	Beta-thal trait	Used within 3 days	T9/94	RPMI1640/25 mM Hepes/10% serum Medium change on day 2 and 4; Ht in culture: approximately 5%	Candle jar	No difference in beta-thal trait	↓ in beta-thal trait

(*Continued*)

Table 3.1. Summary of the Effect of Some Malaria-Protective Mutations on Invasion and Growth of *P. falciparum*—Cont'd

Study no.	Reference	Mutation	RBC status time after withdrawal)	*P. falciparum* strain	Growth medium and culture details	Incubation conditions	Invasion (I) and Growth (G)
6	Yuthavong *et al.* (1987)	Beta-thal trait	Used within 24 h	K_1	RPMI1640/ 25 mM Hepes/ 10% serum; Daily medium change; Ht in culture: approximately 50%	Candle jar	Normal multiplication rate after 4 days in culture
7	Senok *et al.* (1997)	Beta-thal trait	Used within 48 h	FC27	RPMI1640/25 mM Hepes/10% serum; Daily medium change; Ht in culture: approximately 4%	5% O_2, 5% CO_2	↓ invasion and re-invasion rate No difference in G (day 1–4), ↓↓ at day 6
8	Udomsangpetch *et al.* (1993)	Beta-thal trait	Used within 1 week	TM267R	RPMI1640/25 mM Hepes/10% serum; Ht in culture: 3%	20% O_2, 5% CO_2	↓↓ multiplication rate No difference in G (↓ in 4/28 cases).
9	Pattanapanyasat *et al.* (1999)	Heterozygous alpha-thal	Used within 1 week	TM267TR	RPMI1640/25 mM Hepes/10% serum; Ht in culture: 3%	20% O_2, 5% CO_2	No difference in G ↓↓ multiplication rate
10	Luzzi *et al.* (1990, 1991)	Heterozygous alpha^0-thal	Used within 48 h	Palo Alto T9/96	RPMI1640/25 mM Hepes/10% serum; Ht in culture: approximately 3%	Sealed jar 1% O_2, 5% CO_2, 94% N_2	No difference in I and G

					approximately 50%			
11	Brockelman *et al.* (1987)	HbH	Used within 3 days	T9/94	RPMI1640/25 mM Hepes /10% serum; medium change on day 2 and 4 Ht in culture: approximately 5%	Candle jar	↓↓ **I**	No G
12	Pattanapanyasat *et al.* (1999)	HbH	Used within 1 week	TM267TR	RPMI1640/25 mM Hepes /10% serum; daily medium change; Ht in culture: approximately 4%	20% O_2, 5% CO_2	No difference in **I** ↓↓ multiplication rate	
13	Ifediba (1985)	HbH	Used within 4 h	NF-77	RPMI1640/25 mM Hepes /10% serum/ 50 µg/mL hypoxanthine; Daily medium change; Ht in culture: 8%	Candle jar	↓ **I**	↓↓ G

Comments (the numbers refer to study number in the table)
1. Carefully conducted study biased by the generic questionable significance of in vitro data. Study performed culturing parasites in the more appropriate medium 199, better than RPMI1640, due to the presence of hypoxanthine. Authors do not find evidence of increased sickling in parasitized Hb AS RBCs. Slight reduction in rate of invasion under low oxygen tension but show marked retardation of parasite development in Hb AS RBCs. Statistical significance of results with Hb AS RBCs questioned by Laser and Klein (*Nature*, **280,** [1979] 613–614): "The data given in their Table 3.1 do not, in our view, support the hypothesis that low oxygen tension impedes the rate of invasion and development by merozoites of *P. falciparum* in erythrocytes containing Hb S".
2. Results biased by the generic questionable significance of in vitro data obtained using RPMI1640 without supplements and old RBCs. The Author (see Friedman, 1978) shows that immature parasite forms maintained in continuous culture did not develop in Hb AS cells under reduced O_2 tension. Validity of result is biased because parasite immature forms are mostly found in well oxygenated peripheral blood.
6. Inhibition becoming patent only after long cultivation periods, and absence of inhibition in case of sub-cultivation, points at indirect, unspecific effects derived from lack of nutrients/accumulation of toxic waste products, higher in thalassemia-RBCs because of their larger numbers (microcytosis effect!)
7. Clear inhibition of multiplication ratio (parasitemia at day 4/parasitemia at beginning) in double heterozygotes beta-thalassemia/Hb E. Double heterozygotes are more frequent in Thailand compared to simple thalassemia- or Hb-E traits, and therefore possibly more protective. However, clinical studies showing better protection of double vs mono-heterozygotes are missing.
13. Only four black patients with different genotypes and mild hematological symptoms examined. The Asian form of HbH (see studies No. 11 and 12) is usually accompanied by severe hematological symptoms. HbH disease has very low overall incidence and evidently little selective advantage. No evidence of HbH protection against severe malaria *in vivo*. Parasite growth in normal controls declining after 4 days in culture, pointing at lack in nutrients, as probably occurring at the very high Ht and inadequate culture medium adopted in the study

higher numbers of mutant cells compared to nonmutant controls. In addition, thalassemic RBCs have 3.5-fold higher glucose consumption (Ting *et al.*, 1994) and a reduced transport of adenosine (Mynt *et al.*, 1993). Likewise, due to increased endogenous oxidative stress, lower NAD redox potential, and increased glycolysis (Al-Ali, 2002), Hb AS RBCs may also have increased metabolic rate. In the following, a few examples are considered in more detail.

4.1. Sickle-Cell Anemia

In a study performed with Hb AS RBCs, a modest difference in invasion was observed in mutant RBC kept at 5% oxygen, corresponding to the partial pressure of 35 mmHg present in venous blood (Pasvol *et al.*, 1978; Pasvol, 1980). Recalculation of data for parasites grown in 5% CO_2 in air indicated no difference in growth in Hb AS RBCs. In the same study, early parasite forms were allowed to develop to maturity at low oxygen/5% CO_2 in an attempt to mimic conditions in deep venous blood. Under these conditions, growth of mature parasites was retarded in the Hb AS cells, although the recalculated total number of mature forms was very similar in Hb AS and Hb SS cells (Laser and Klein, 1979). Extrapolation of these results to *in vivo* conditions was further limited by the small number (3) of experiments. (Pasvol *et al.*, 1978; Pasvol, 1980).

4.2. Beta-Thalassemia

Impaired parasite growth in heterozygous beta-thalassemia RBCs *in vitro* has been described by some authors (Brockelman *et al.*, 1987; Kaminsky *et al.*, 1986), while others found no difference (Yuthavong *et al.*, 1987). The significance of some results showing inhibition could not be assessed because scanty description of culture conditions did not allow comparison to data obtained by other studies (Kaminsky *et al.*, 1986). Brockelman's data are well documented (Brockelman *et al.*, 1987). However, their results are also open to criticism. These authors did not see any difference during 5 days in culture in beta-thalassemia/Hb E RBCs if the cultivation was performed in RPMI1640 medium. Differences became significant by lowering overall amino acid levels in the medium, in the attempt to mimic lower hemoglobin levels in the mutant RBCs (Brockelman *et al.*, 1987). These data are interesting because they stress the growth-limiting role of amino acid of the medium and show that indeed mutant RBCs were more sensitive to limiting amino acid levels than normal controls. Senok *et al.* (1997) analyzed parasite invasion and growth in density (age)-separated RBCs over 144 h (i.e., three growth cycles). After 144-h incubation, parasitemia in cultures was lower in nonseparated beta-thalassemia samples, and differences were more pronounced in the less dense fraction, where younger, metabolically more active RBCs were localized. Differences were much smaller in the old RBC fraction. Moreover, no or irrelevant differences were noted in nonseparated RBCs until 96 h of culture. A schizont maturation arrest and morphological alterations of late trophozoite stages were also observed in beta-thalassemia samples. Recently, Pattanapanyasat *et al.* (1999) have shown that normal and thalassemic RBCs were equally susceptible to invasion and growth during the first invasion, while parasite growth was significantly reduced in the successive cycles. Both Senok's and Pattanapanyasat's studies were run at relatively high hematocrit (3–4%) over long period of times with the questionable RPMI1640 medium. Therefore, it cannot be excluded that nutrient limitation at late time points, combined with higher metabolic rates in mutant RBCs (see further) may explain the observations.

4.3. Alpha-Thalassemia

Two important studies by Luzzi *et al.* (1990, 1991) have shown no difference in parasite growth in alpha-thalassemia, and different antibody density on the RBC surface, compatible with enhanced immune removal of the mutant parasitized RBCs. Other studies performed by the same groups discussed before with different types of alpha-thalassemia gave comparable results. In those studies, most severe conditions, such as in hemoglobin H disease, were associated with remarkable impairment of parasite growth, especially after longer periods of cultivation. In conclusion, analysis of some representative studies showing impaired parasite growth in RBCs heterozygous for protective mutations (alpha-thalassemia, beta-thalassemia and sickle-cell disease) indicates that this effect is compatible with nutrient limitations affecting parasite growth and therefore possibly of artifactual origin. This conclusion seems plausible because all studies were conducted with media devoid of hypoxanthine and adenine and containing low concentrations of a number of amino acids, at relatively high hematocrit and for prolonged period of times. Mutations considered here are particularly sensitive to nutrient deprivation because they have lower intracellular hemoglobin and thus less endogenous amino acid, and higher metabolic demands due to permanent oxidant stress related to unpaired globin chains, sickle hemoglobin, and high membrane free iron (Weatherall 1998; Hebbel, 1990). In addition and perhaps most importantly, nonparasitized Hb AS-RBCs and thalassemia-RBCs are dehydrated and/or microcytic (Weatherall 1998; Hebbel, 1990). Thus, the number (and activity) of metabolically active elements per unit of blood volume is remarkably large compared to normocytes, further increasing metabolic activity that is proportional to total cell surface.

4.4. Glucose-6-Phosphate Dehydrogenase Deficiency

Deficiency of this enzyme is very widespread. Estimates based on population studies and older population data indicate 400,000 million of deficient hemi-heterozygous subjects worldwide, a likely underestimate of the real occurrence. Most widespread variant alleles with polymorphic frequencies are the Mediterranean variant (1–5% residual activity), frequent in Mediterranean countries and in the Middle East; the A^- variant (approximately 5–50% residual activity), frequent in Africa and America; the Orissa variant and the India–Pakistan Mediterranean sub-variant (low residual activities), frequent in India and Pakistan; the Canton, Taiwan and Mediterranean variants (low residual activities), frequent in China and Asia (Roberts *et al*., 2004; Greene, 1993). Marked differences in residual activity of variants should be taken into consideration in discussing their protective efficiency. In general, subjects with variants with high residual activity are expected to be less protected, because less prone to oxidants, compared with subjects with the higher-protective Mediterranean variant, that are much more sensitive toward oxidants. This contention cannot be proven now, though because areas where the Mediterranean variant is frequent are presently malaria-free.

In vitro parasite growth studies have yielded conflicting results. Taking into account homogeneous studies performed with the Mediterranean variant, a marked reduction in parasitemia in cultures from hemizygotes and heterozygotes was observed by Roth *et al.* (1983) 5 days after inoculation. Luzzatto *et al.* (1983), by contrast, observed similar growth in the first and a 40% reduction in the second schizogonic cycle. More recent data performed with five different strains during three schizogonic cycles showed no difference in growth, provided the RBCs were fresh, the hematocrit level of the culture low and glucose in culture kept high and constant (Cappadoro *et al*., 1998). Difference in growth was noted by the same authors when older RBCs were used. It appears that the same explanation for inhibition of parasite growth in sickle-cell trait RBCs may apply to G6PD-deficiency, and it seems plausible that nutrient deprivation, or cell damage due to extended storage coupled with increased sensitivity of mutant RBCs to oxidant stress may be responsible for impaired parasite growth.

5. Modifications in the RBC Membrane Elicited by the Developing Parasite Induce Phagocytosis

5.1. Similarity of Ring Phagocytosis to Phagocytosis of Normal Senescent or Oxidatively Stressed RBCs

P. falciparum-parasitized RBCs are progressively transformed into non-self cells, opsonized and phagocytosed by circulating and resident phagocytes. At ring stage, phagocytosis was modest and almost totally mediated by complement deposition and recognition by complement receptor type 1 (CR1) on the phagocyte. At later stages of parasite maturation, phagocytosis was strongly increased and the role of complement in phagocytic recognition was reduced (Turrini *et al.*, 1992). Data obtained with nonparasitized RBCs have shown that binding of oxidative denaturation products of hemoglobin (hemichromes) powerfully induced band 3 aggregation, expression of neoantigens, deposition of IgG and complement, and phagocytosis (Low *et al.*, 1985; Lutz, 1990; Mannu *et al.*, 1995; Turrini *et al.*, 1991; Cappellini *et al.*, 1999). Phagocytosis was largely complement-dependent and mediated by CR1 on the phagocyte (Mannu *et al.*, 1995; Turrini *et al.*, 1991). It has been shown that hemichromes were increasingly generated during parasite development and bound stage-dependently to the RBC membrane (Giribaldi *et al.*, 2001). In parallel, increasing amounts of high-molecular weight clusters of aggregated band 3 were formed. Those clusters were constantly associated with complement and autologous IgG (Giribaldi *et al.*, 2001). The parasite-induced modifications were of oxidative nature, and aggregation of band 3 and deposition of opsonins could be largely reverted by cultivating the parasite in the presence of beta-mercaptoethanol (Giribaldi *et al.*, 2001). It has also been shown that growth of *P. falciparum* induced hemichrome formation in the host RBCs. Hemichromes were isolated from large membrane protein aggregates that also contained band 3, IgG and the complement C3c fragment, the stable derivative of C3b. This finding suggests that hemichromes bound to its specific, high-affinity binding site on the cytoplasmic domain of band 3 and induced band 3 aggregation, in agreement with literature data (Low *et al.*, 1985) and previous observations by our group (Cappellini *et al.*, 1999). Indeed, present data are similar to observations performed in thalassaemic RBCs, where membrane protein aggregates also contained large amounts of IgG and C3 complement fragments (Mannu *et al.*, 1995), and where deposition of IgG and complement C3c, and phagocytic susceptibility were correlated to the amount of membrane-bound hemichromes (Cappellini *et al.*, 1999). During the parasite development, the amount of hemichromes bound to the membrane progressively increased and paralleled the amount of aggregated band 3 and deposited IgG and C3c suggesting a causal relationship between hemichrome formation and band 3 aggregation (Giribaldi *et al.*, 2001). The involvement of oxidative events in the formation of the antigenic membrane aggregate was substantiated by the inhibition of their formation when the parasites were cultivated in presence of permeant beta-mercaptoethanol, known to be a powerful reductant and oxidant scavenger. Immunoprecipitation studies further substantiated the connection between membrane-bound IgG and band 3. Immunoprecipitated band 3 was oxidatively cross-linked, as no electrophoretic bands were obtained in absence of reductive pretreatment, and migrated as a sharp band, which corresponded to a fast-migrating, underglycosylated band 3 component and was absent in the unreduced membranes from parasitized RBCs. The latter results indicated that membrane-bound IgGs were linked to band 3 and showed that apparently underglycosylated band 3 is preferentially incorporated into the membrane aggregates (Giribaldi *et al.*, 2001). To characterize the band 3 modifications that lead to its recognition by naturally occurring antibodies, the IgG bound to the surface of parasitized RBCs were eluted and challenged with membrane proteins extracted from nonparasitized RBCs, parasitized RBCs and chemically treated RBCs (diamide and zinc-BS3 treatments) to obtain well-defined oxidative (diamide) and nonoxidative (zinc-BS3) band-3 aggregation. This approach further demonstrated that surface-bound IgG had affinity to oxidatively aggregated band 3. These data partially confirmed and expanded previous observations done in malaria-parasitized human RBCs erythrocytes (Winograd and Sherman, 1989a, 1989b; Winograd *et al.*, 1987), showing the formation and exposure of two neoantigens: a 85-kDa antigen and a >250-kDa high-molecular weight antigen, both structurally related to band 3 and recognized by naturally occurring anti-band 3 autoantibodies. Stage-dependent increases in RBC-bound C3c complement fragment were consistently observed here in agreement with studies showing complement activation by the surface of *P. falciparum*-parasitized RBCs and complement-mediated phagocytosis of ring-parasitized human RBC (Turrini *et al.*, 1992). It was suggested that parasite development induces, in sequence, deposition of hemichromes and oxidative aggregation of band 3; deposition of autologous antiband 3 IgG and complement; and final recognition by phagocytes. This sequence is quite similar to that observed in normally senescent RBCs, artificially modified RBCs or pathologic RBCs, most notably sickle and thalassemic RBCs (Low *et al.*, 1985; Lutz, 1990; Mannu *et al.*, 1995;

Turrini *et al.*, 1991; Cappellini *et al.*, 1999; Kannan *et al.*, 1988). The mechanism of phagocytic removal of parasitized RBCs was progressively enhanced as parasite maturation proceeded and attained the highest values with trophozoite-parasitized and schizont-parasitized RBCs.

6. Enhanced Phagocytosis of Ring Forms as a Model of Protection for Widespread RBC Mutations

6.1 Membrane Binding of Hemichromes, Autologous IgG, Complement C3c Fragment; Aggregated Band 3 and Phagocytosis in Nonparasitized and Ring-Parasitized Normal and Mutant RBCs

The levels of indicators of membrane damage that induce phagocytosis, such as membrane-bound hemichrome, IgG and complement C3c, and aggregated band 3, are already increased in nonparasitized Hb AS, beta-thalassemia trait and HbH RBCs (see Figure 3.1). The highest level of damage was noted in nonparasitized Hb AS and HbH RBCs while alpha-thalassemia trait RBCs was indistinguishable from normal controls. The development of the parasite in those mutant RBCs further increased the levels of above indicators of membrane damage in ring-parasitized mutant RBCs with the exception of alpha-thalassemia trait RBCs. Phagocytosis was also very remarkably increased in ring-parasitized RBCs in all mutations except alpha-thalassemia trait (Figure 3.2). Heme-containing compounds, identified as hemichromes from their

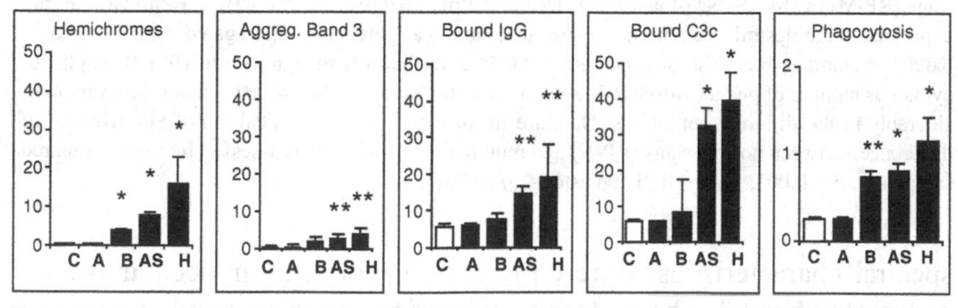

Figure 3.1. Membrane-Bound Hemichromes, Autologous IgG and Complement C3c Fragment; Aggregated Band 3, and Phagocytosis in Nonparasitized Normal and Mutant RBCs. Parameters were measured in nonparasitized normal control (C), alpha-thalassemia trait (A), beta-thalassemia trait (B), Hb AS (AS) and HbH (H) RBCs. Hemichromes are expressed in nmol/mL membranes; aggregated band 3 as percentage aggregated band 3 over total band 3; membrane-bound autologous IgG and C3c as milliabsorbance units/min/10^7 RBCs; phagocytosis as number of phagocytosed RBCs per monocyte. Data are mean values ± SD (vertical bars). Numbers of separate experiments each performed with a different donor were: Hb AS, 11; beta-thalassemia trait, 12; alpha-thalassemia trait, 4; HbH, 5. Significance of differences between nonparasitized normal and mutant RBCs was assessed by *t*-test for paired samples. *, $p < 0.001$; **, $p < 0.01$; no asterisk, $p > 0.05$.

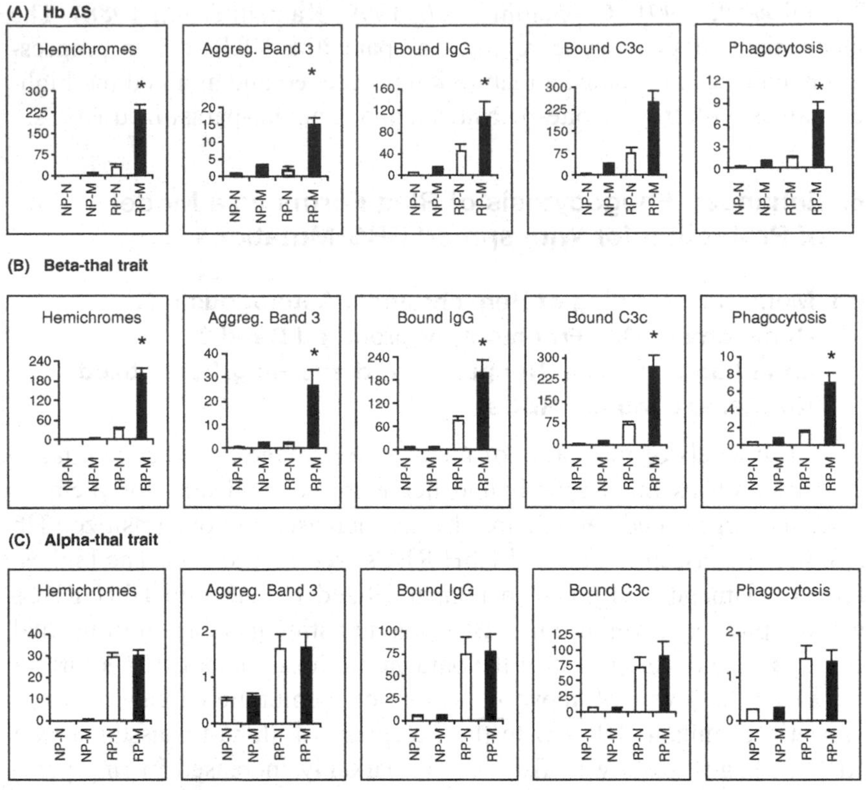

Figure 3.2. Membrane-Bound Hemichromes, Autologous IgG and Complement C3c Fragment; Aggregated Band 3 and Phagocytosis in Nonparasitized Normal and Mutant RBCs, and in Ring-Parasitized Normal and Mutant RBCs. Parameters were measured in nonparasitized normal controls (NP-N), nonparasitized mutant controls (NP-M), normal rings (RP-N) and mutant rings (RP-M) in Hb AS, beta-thalassemia trait and alpha-thalassemia trait RBCs. Hemichromes are expressed in nmoles/mL membranes; aggregated band 3 as percentage aggregated band 3 over total band 3; membrane-bound autologous IgG and C3c as milliabsorbance units/min/10^7 RBCs; phagocytosis as number of phagocytosed RBCs per monocyte. Note that the ordinate values may vary considerably in the different conditions. Data are mean values ± SD (vertical bars). Significance of differences between normal rings (RP-N) and mutant (RP-M) rings was assessed by *t*-test for paired samples. *, $p < 0.001$; **, $p < 0.01$; no asterisk, $p > 0.05$.

spectral characteristics, were co-localized with high-molecular weight aggregated band 3. The co-localization of hemichromes and aggregated band 3 is considered as indication of hemichrome-induced clustering of band 3. The number of phagocytosed Hb AS, beta-thalassemia trait and Hb H rings was close to 10 ingested RBCs per monocyte, the maximal erythrophagocytic capacity of adherent human monocytes. The role of complement as opsonin was tested in ring-parasitized beta-thalassemia trait RBCs. Abrogation of C3-mediated phagocytosis by blockage of CR1 receptor on monocytes reduced phagocytosis of ring-parasitized beta-thalassemia trait and normal RBCs by approximately 80 and 94%, respectively, in agreement with previous observations (Turrini *et al.*, 1992). Decrease of C3-mediated phagocytosis was less pronounced in trophozoites (46 and 43% reduction in trophozoite-parasitized beta-thalassemia trait and normal RBCs, respectively), indicating that IgG- and scavenger receptors may play a more significant role in phagocytosis of mature parasite forms (Turrini *et al.*, 1992). Latter data are in agreement with observations in trophozoite-parasitized G6PD-deficient RBCs. The oxidative origin of membrane damage was underscored by a partial reversion of hemichrome formation, band 3 aggregation, and deposition of removal markers in normal and mutant RBCs cultivated in presence of 100 μmol/L beta-mercaptoethanol (Giribaldi *et al.*, 2001). Taken together, these results indicate that membrane damage ultimately generated by the interaction of abnormal hemoglobins and the parasite and likely responsible for enhanced phagocytosis of ring-parasitized mutant RBCs, was very similar to ring-parasitized G6PD-deficient RBCs and to senescent or oxidatively damaged nonparasitized RBCs (Lutz, 1990; Giribaldi *et al.*, 2001).

6.2 Why are Rings Developing in Beta-Thalassemia, Sickle-Cell Trait, HbH, and G6PD-deficient RBCs Phagocytosed more Intensely than Rings Developing in Normal RBCs?

The molecular nature of Hb AS, beta-thalassemia and alpha-thalassemia trait, and G6PD-deficiency is different. However, in all cases affected RBCs show increased production of reactive oxygen species (ROS), due to intrinsic characteristics of Hb S in Hb AS, to unpaired globin chains in the thalassemias and to defective antioxidant defense in G6PD deficiency. As detailed before, oxidative events leading to enhanced phagocytosis of pathological, artificially modified, and ring-parasitized RBCs are in sequence: increased denaturation of Hb and formation of reversible and irreversible hemichromes; membrane binding of hemichromes to the cytoplasmic domain of band 3; membrane deposition of free iron; hemichrome-induced and free-iron-induced aggregation of band 3; limited activation of the alternative complement pathway; and deposition of antibodies and complement C3c fragments.

Why are rings developing in beta-thalassemia, sickle-cell trait, Hb H, and G6PD-deficient RBCs phagocytosed more intensely than ring-parasitized normal RBCs? Based on our observations, oxidative membrane damage, deposition of opsonins and phagocytosis were higher in ring-parasitized mutant RBCs compared to "normal" rings. This is explained by the fact that rings developing in mutant RBCs are subjected to a double oxidative stress: a first one exerted by the developing parasite and a second specifically connected with the mutation and additional to the first one. Phagocytosis was similar in trophozoites grown in normal and mutant RBCs, as observed previously in trophozoites grown in G6PD-deficient RBCs. Damages inflicted by mature parasite forms are very profound and overshadow the baseline differences in normal and mutant RBC. In trophozoites, a larger share of phagocytic recognition does not depend on band 3 aggregation but relies on exposure of other molecules. For example, exposure of phosphatidylserine (PS), an indicator of loss of phospholipid asymmetry in the RBC membrane, was remarkable in both mutant and nonmutant trophozoite-parasitized RBCs (Figure 3.3). Interestingly, hemichromes were significantly increased in trophozoites grown in sickle- and beta thalassemia-trait RBCs, whereas aggregated band 3 and phagocytosis did not further increase and were similar to trophozoites grown in normal RBCs. Possible reasons may reside in the limited amount of mobile band 3 in the RBC membrane and in the physical limit of 10 RBCs ingested per monocyte.

The evident phenotypic similarities among many protective mutations, irrespective of the molecular nature of the underlying defect, have been noted in other studies (Destro-Bisol *et al.*, 1996). Not surprisingly, higher levels of bound antibodies and more intense phagocytosis have been described in parasitized thalassemic and other mutant RBCs. Thus, enhanced removal of parasitized mutant RBCs by the host's immune system has been suggested as the mechanism underlying resistance. The model presented here is also based on preferential immunological removal of parasitized mutant RBCs. However, it has the distinctive feature that only phagocytosis of ring-parasitized mutant RBCs is selectively enhanced, while phagocytosis of trophozoites is very high but quite similar in normal and mutant cells. This model of resistance based on enhanced

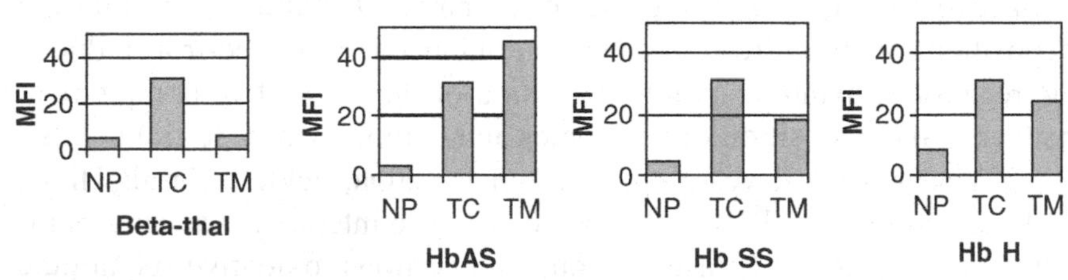

Figure 3.3. Loss of Phospholipid Asymmetry in Trophozoite-Parasitized Normal and Mutant RBCs. NP, nonparasitized normal controls; TC, trophozoite-parasitized normal RBCs; TM, trophozoite-parasitized mutant RBCs. MFI, mean fluorescence intensity of Annexin-V labeled RBCs measured by cytofluorimetry.

phagocytosis of ring-parasitized mutant RBCs was originally proposed for G6PD-deficiency (Cappadoro *et al.*, 1998) and successively extended to include Hb AS, beta-thalassemia trait, and Hb H (Ayi *et al.*, 2004). Hb H has no selective value against malaria, since it is accompanied by severe hematologic symptoms. It was been added to the study to show that the different behavior of beta-thalassemia trait (enhanced ring phagocytosis) and alpha-thalassemia trait (no enhancement of ring phagocytosis) resides in the different amount of oxidative damage between the two thalassemic conditions. As soon as the damage increased, as in HbH disease, ring-phagocytosis was enhanced as well.

6.3 What is the Evidence that Ring-Parasitized Mutant RBCs are Also Preferentially Removed *In Vivo*?

Supporting evidence that preferential phagocytosis of mutant rings is occurring *in vivo* can be only provided by mutations with phenotipically distinct mutant and normal RBC populations in the same subject, such as G6PD-deficient female heterozygotes that have a mosaic RBC population, one G6PD-deficient and one normal in the same subject. Indeed, in malarious G6PD-deficient heterozygous females, a prevalence of parasitized normal RBCs vastly in excess over parasitized deficient RBCs was described (Luzzatto *et al.*, 1969). This unbalance is best explained by the selective removal ("suicidal infection") of ring-parasitized deficient RBCs from circulation. A similar behavior was observed in malarious Hb AS subjects. In Hb AS, all RBCs are genetically equivalent and contain both Hb A and Hb S; and, *a priori*, have the same chance of being parasitized and to sickle. However, ring-parasitized Hb AS RBC were found to sickle approximately six times as readily as non-parasitized cells, generating a double, phenotypically different population—parasitized sickled cells, more prone to be phagocytosed, and nonparasitized nonsickled cells (Luzzatto *et al.*, 1970). Indeed, Hb AS malarious patients showed a remarkable prevalence of nonsickled vs sickled ring-parasitized RBCs in peripheral blood, an unbalanced situation comparable to G6PD-deficient malarious heterozygotes (Mackey and Vivarelli, 1954). Also malarious SAO heterozygotes, a mutation not considered in detail here, have a double population of ovalocytic and normal RBCs and displayed selective disappearance of ovalocytic RBCs, in accordance with their preferential phagocytosis (O'Donnell *et al.*, 1998). SAO is caused by a band 3 mutation due to a 9-aminoacid deletion in the N-terminal of band 3, producing very rigid RBCs with higher amounts of immobile, microaggregated band 3 (Tilley *et al.*, 1991; Liu *et al.*, 1995). It is likely that the presence of the parasite may further enhance the increased baseline values of aggregated band 3, reproducing by a partially different mechanism the situation observed here in ring-parasitized sickle-cell and beta-thalassemia trait RBCs.

6.4 Why is Preferential Removal of Ring-Forms Advantageous to the Host?

Enhanced and preferential phagocytosis of ring-parasitized mutant RBCs may be advantageous to the host in several ways. A first advantage is reduction in parasite growth and parasite density, observed in patients with Hb AS and beta-thalassemia trait but not in patients with alpha-thalassemia trait. Second, phagocytosed ring-forms are digested rapidly by monocytes, and the process repeated without loss of efficiency. By contrast, more mature forms of the parasite, although actively phagocytosed, severely affect important functions of the monocyte, such as the ability to repeat the phagocytic process. Adverse effects elicited by ingestion of hemozoin-containing parasite forms include enhanced production of inflammatory cytokines, inability to kill ingested bacteria, to perform repeated cycles of phagocytosis, to express class II and other membrane antigens upon interferon gamma stimulation, and to correctly process antigens (Schwarzer *et al.*, 1992; Scorza *et al.*, 1999). Endothelial functions are also impaired by hemozoin, and adhesion of mature parasite forms to dendritic cells and macrophages downregulates innate and acquired immune responses (Urban and Roberts, 2002). Third, lower numbers of trophozoites and schizonts, that adhere to endothelia in several important organs (lungs, kidneys, brain, bone marrow, and placenta) and provoke severe symptoms (e.g., cerebral malaria, placental malaria, possibly dyserythropoiesis, and respiratory distress) may also lead to less severe disease and lower mortality (Ho and White, 1999). Finally, phagocytosis of ring-parasitized normal and mutant RBCs was accompanied by a very modest oxidative burst by monocytes (not shown). Also, complement-mediated phagocytosis was shown to induce a reduced cytokine output by the phagocytic cells.

6.5 Why are Rings Developing in Alpha-Thalassemia Very Similar to Controls?

Alpha-thalassemia trait rings were not different from nonmutant rings as to hemichrome levels, deposition of removal markers, and phagocytosis and were clearly distinct from beta-thalassemia and Hb H rings. All thalassemic syndromes are characterized by unbalanced globin

Table 3.2. Comparison Between Conditions Present *In Vivo* and in Continuous Culture *In Vitro* of *P. falciparum*

	Cells suspended in	Ht (%)	O_2[a]	CO_2	Osmotic pressure[b]	Extracellular pH	Flow shear stress	Other cells types
In vivo	Plasma	30–45	Oscillating (High-low)	Low and oscillating	Variable (up to 1000 mOsm)	7.45	Present	Present
In vitro	Culture medium	3–8	Nonoscillating	High and constant	Constant (300 Osm)	7.00–7.20	Absent	Absent

[a] Oxygen partial pressure oscillates between 100 mmHg in arterial blood and 20–40 mmHg in venous blood approximately six times per minute in the general circulation.
[b] Osmotic pressure in the peritubular capillaries of kidney where RBCs may circulate for as long as 60–90 s.

chain synthesis and membrane deposition of excess unpaired alpha-chains or beta-chains in beta-thalassemia and alpha-thalassemia trait, respectively. However, pathophysiological consequences are very different. A number of studies (see Schrier *et al.*, 1994 for review) and present data indicate that membrane deposition of alpha-chains in beta-thalassemia inflicts distinctly more severe damages, compared to deposition of beta-chains. For example, mechanical stability of membranes was increased in alpha-thalassemia and markedly decreased in beta-thalassemia; and alpha-chains but not beta-chains bound band 3 cytoplasmic domains with high affinity and positive cooperativity (Schrier, 1994). The modest degree of alterations observed in alpha-trait nonparasitized and ring-parasitized RBCs would exclude the same pattern of resistance to be operating *in vivo*. There is no doubt that alpha-thalassemia trait is protective, as shown by its coincidental presence with past or present *falciparum* malaria, as in the Tharu population in Nepal, in Melanesia, and Africa (Flint *et al.*, 1986, 1993; Yenchitsomanus *et al.*, 1986; Allen *et al.*, 1997; Modiano *et al.*, 1991). However, careful studies by the Oxford group have shown that significantly lower prevalence of severe malaria in thalassemia-trait children was not accompanied by any difference in parasite density or mortality due to malaria complications (Allen *et al.*, 1997). These studies found evidence for raised rather than reduced incidence of mild malaria in children carrying alpha-thalassemia (Williams *et al.*, 1996). Most likely, enhanced ring-phagocytosis is not operating in alpha-thalassemia trait, whereby the suggestion that alpha-thalassemia protects by predisposing to mild malaria in early life and provoking a cross-vaccination by coincidental co-infection with *P. vivax* may offer an interesting alternative explanation.

7. Conclusions

Diffusion of *falciparum* malaria, most probably starting with the formation of extensive human aggregates, which accompanied the agricultural revolution, has posed a very powerful challenge to the plasticity of defensive responses of the host. It is now a commonplace notion that malaria had a profound impact on recent human evolution. The "malaria hypothesis," commonly associated with the name of J.B.S. Haldane, proposes that widespread human genetic polymorphisms, especially those affecting RBCs, have been selected to high frequencies because they protect against the most dangerous or deadly clinical manifestations of malaria. Geographical distribution and clinical studies have provided evidence that sickle-cell trait, heterozygous alpha and beta-thalassemias, Southeast Asian ovalocytosis, Hb C and Hb E, and hemizygous/

heterozygous G6PD-deficiency are protective RBC mutations. The predominant explanation found in the literature for resistance provided by above conditions is based on *in vitro* parasite cultivation studies claiming that the mutant RBCs do not offer optimal conditions for either invasion or parasite development. We have discussed supportive evidence provided by many of those studies and found them methodologically biased. Often, a first bias resided in the impossibility to reproduce the very complex *in vivo* situation with the simplified experimental setup adopted *in vitro*. A second bias resulted from the generalized use of inadequate culture media low in reduced glutathione and devoid of hypoxanthine and adenine, both essential for parasite growth. The limiting level of essential components of culture media, such as essential amino acids, could critically affect parasite development because thalassemic and Hb AS RBCs have a higher glucose consumption and an increased metabolic rate. We have proposed an alternative explanation based on enhanced phagocytosis of early parasite forms (ring forms) when the parasite developed in RBCs affected by the mutations. The molecular nature of Hb AS, beta-thalassemia and alpha-thalassemia trait, and G6PD-deficiency is different. However, mutant RBCs produce increased amounts of ROS, due to the intrinsic characteristics of mutant hemoglobins, the presence of unpaired globin chains in thalassemias or the defective antioxidant defense in G6PD deficiency. Indeed, baseline levels of phagocytic markers, such as membrane-bound hemichrome, IgG and complement C3c, and aggregated band 3, were enhanced in nonparasitized RBCs affected by protective mutations with the exception of alpha-thalassemia. We have recently shown that the parasite developing in mutant RBCs further enhanced the baseline oxidative changes and specifically enhanced ring phagocytosis. In parasitized mutant RBCs, membrane changes induced at ring stage were very similar to changes observed in nonparasitized senescent or oxidatively damaged RBCs. The differences in phagocytosis of mutant-parasitized vs nonmutant-parasitized RBCs were ring-specific and vanished at trophozoite stage. By contrast, in all considered mutations, invasion and parasite development was not different in normal vs mutant RBCs. We suggest that enhanced and preferential phagocytosis of ring-parasitized RBCs may be advantageous to the host because it reduced parasite growth and parasite density and lowered the number of late-stage parasites adherent to venular endothelia. Mature forms of the parasite are intensely phagocytosed and severely affect important functions of the monocyte. They enhance production of inflammatory cytokines, reduce the killing of ingested bacteria, lower expression of class II and other membrane antigens upon interferon gamma stimulation, and inhibit antigen processing

(Schwarzer *et al.*, 1992; Scorza *et al.*, 1999). Endothelial functions are also impaired by phagocytosis of mature forms, and their adhesion to dendritic cells and macrophages downregulates innate and acquired immune responses (Urban and Roberts, 2002). Third, lower numbers of trophozoites and schizonts that adhere to endothelia in several important organs (lungs, kidneys, brain, bone marrow, and placenta) and provoke severe symptoms (e.g., cerebral malaria, placental malaria, possibly dyserythropoiesis, and respiratory distress) may also lead to less severe disease and lower mortality (Ho and White, 1999).

References

Agarwal, A., Guindo, A., Cissoko Y. *et al.* (2000). Hemoglobin C associated with protection from severe malaria in the Dogon of Mali, a West African population with a low prevalence of hemoglobin S. *Blood*, **96,** 2358–2363.

Akide-Ndunge, O., Ayi, K., and Arese, P. (2003). The Haldane malaria hypothesis: Facts, artifacts, and a prophecy. *Redox Rep.*, **8,** 311–317.

Al-Ali, A.K. (2002). Pyridine nucleotide redox potential in erythrocytes of Saudi subjects with sickle-cell disease. *Acta Haematol.*, **108,** 19–22.

Allen, S.J., O'Donnell, A., Alexander, N.D. *et al.* (1997). Alpha$^+$-thalassemia protects children against disease caused by other infections as well as malaria. *Proc. Natl. Acad. Sci. USA*. **94,** 14736–14741.

Allen, S.J., O'Donnell, A., Alexander, N.D. *et al.* (1999). Prevention of cerebral malaria in children in Papua New Guinea by southeast Asian ovalocytosis band 3. *Am. J. Trop. Med. Hyg.*, **60,** 1056–1060.

Allison, A.C. (1954). Protection afforded by sickle-cell trait against subtertian malaria infection. *Br. Med. J.*, **i,** 290–294.

Allison, A.C. (1956). The sickle-cell and haemoglobin C genes in some African populations. *Ann. Hum. Genet.* **21,** 67–89.

Allison, A.C. (1964). Polymorphism and natural selection in human populations. *Cold Spring Harbor Symp. Quant. Biol.*, **29,** 137–149.

Allison, A.C. and Clyde, D.F. (1961). Malaria in African children with deficient glucose-6-phosphate dehydrogenase. *Br. Med. J.*, **i,** 1346–1348.

Ayi, K., Turrini, F., Piga, A., and Arese, P. (2004). Enhanced phagocytosis of ring-parasitized mutant erythrocytes: a common mechanism that may explain protection against *falciparum* malaria in sickle trait and beta-thalassemia trait. *Blood*, **104**, 3364 - 3371

Beet, E.A. (1947). Sickle-cell disease in Northern Rhodesia. *E. Afr. Med. J.*, **24,** 212–222.

Bianco-Silvestroni, I. (2002) *Storia della microcitemia in Italia*. G. Fioriti Editore, Roma.

Bookchin, R.M. and Lew, V.L. (2002). Sickle red cell dehydration: Mechanisms and interventions. *Curr. Opin. Hematol.*, **9,** 107–110.

Brockelman, C.R., Wongstattayanont, B., Tan-Ariya, P., and Fucharoen, S. (1987). Thalassemic erythrocytes inhibit in vitro growth of *Plasmodium falciparum*. *J. Clin. Microbiol.*, **25,** 56–60.

Bunyaratvej, A., Butthep, P., Yuthavong, Y. *et al.* (1986). Increased phagocytosis of *Plasmodium falciparum*: Infected erythrocytes with haemoglobin E by peripheral blood monocytes. *Acta Haematol.*, **76,** 155–158.

Cappadoro, M., Giribaldi, G., O'Brien, E. *et al.* (1998). Early phagocytosis of glucose-6-phosphate dehydrogenase (G6PD)-deficient erythrocytes parasitized by *Plasmodium falciparum* may explain malaria protection in G6PD deficiency. *Blood*, **92,** 2527–2534.

Cappellini, M.D., Tavazzi, D., Duca, L. *et al.* (1999). Metabolic indicators of oxidative stress correlate with haemichrome attachment to membrane, band 3 aggregation and eythrophagocytosis in beta-thalassaemia intermedia. *Br. J. Haematol.*, **104,** 504–512.

Carcassi, U., Ceppellini, R., and Pitzus, F. (1957). Frequenza della talassemia in quattro popolazioni sarde e suoi rapporti con la distribuzione dei gruppi sanguigni e della malaria. *Boll. Ist. Sieroterap. Milan*, **36,** 207–218.

Ceppellini R. (1955). Negative correlation between altitude above see level and incidence of thalassemia in four Sardinian villages. *Cold Spring Harbor Symp. Quant. Biol.*, **20,** 252.

Chotivanich, K., Udomsangpetch, R., Pattanapanyasat, K. *et al.* (2002). Hemoglobin E: a balanced polymorphism protective against high parasitemias and thus severe *P falciparum* malaria. *Blood*, **100,** 1172–1176.

Destro-Bisol, G., Giardina, B., Sansonetti, B., and Spedini, G. (1996). Interaction between oxidized hemoglobin and the cell membrane: A common basis for several falciparum malaria-linked genetic traits. *Yearb. Phys. Anthropol.*, **39,** 137–159.

Flint, J., Harding, R.M., Clegg, J.B., and Boyce, A.J. (1993). Why are some genetic diseases common? Distinguishing selection from other processes by molecular analysis of globin gene variants. *Hum. Genet.*, **91,** 91–117.

Flint, J., Hill, A.V.S. Bowden, D.K., Oppenheimer, S.J. *et al.* (1986). High frequencies of α-thalassaemia are the result of natural selection by malaria. *Nature*, **321,** 744–750.

Foo, L.C., Rekhraj, V., Chiang, G.L., and Mak, J.W. (1992). Ovalocytosis protects against severe malaria parasitaemia in the Malay aborigines. *Am. J. Trop. Med. Hyg.*, **47,** 271–275.

Friedman, M.J. (1978). Erythrocytic mechanism of sickle cell resistance to malaria. *Proc. Natl. Acad. Sci. USA*, **75,** 1994–1997.

Giribaldi, G., Ulliers, D., Mannu, F. *et al.* (2001). Growth of *Plasmodium falciparum* induces stage-dependent haemichrome formation, oxidative aggregation of band 3, membrane deposition of complement and antibodies, and phagocytosis of parasitized erythrocytes. *Br. J. Haematol.*, **113,** 492–499.

Greene, L.S. (1993). G6PD deficiency as protection against *falciparum* malaria: An epidemiologic critique of population and experimental studies, Year. *Phys. Anthropol.*, **36,** 153–178.

Haldane, J.B.S. (1949a). Disease and evolution. *Ric. Sci.*, **19** (Suppl.), 68–76.

Haldane, J.B.S. (1949b). The rate of mutation of human genes. *Hereditas*, **35** (Suppl.), 267–273.

Hebbel, R.P. (1990) The sickle erythrocyte in double jeopardy: Autoxidation and iron decompartmentalization. *Semin. Hematol.*, **27,** 51–69.

Hill, A.V.S. (2001).The genomics and genetics of human infectious disease susceptibility. *Ann. Rev. Genomics Hum. Genet.*, **2,** 373–300.

Ho M., and White N.J. (1999). Molecular mechanisms of cytoadherence in malaria. *Am. J. Physiol.*, **276,** C1231–C1242.

Hutagalung, R., Wilairatana, P., Looareesuwan, S., Brittenham, G.M. *et al.* (2000). Influence of hemoglobin E trait on the antimalarial effect of artemisinin derivatives. *J. Infect. Dis.*, **181,** 1513–1516.

Ifediba, C.T., Stern, A., Ibrahim, A., Rieder, F.R. (1985). *Plasmodium falciparum* in vitro: Diminished growth in haemoglobin H disease erythrocytes. *Blood*, **65,** 452–455.

Jones, T.R. (1997). Quantitative aspects of the relationship between the sickle-cell gene and malaria. *Parasitol. Today*, **13,** 107–111.

Kaminsky, R., Krüger, N., Hempelmann, E., and Bommer, W. (1986). Reduced development of *Plasmodium falciparum* in β-thalassemic erythrocytes. *Z. Parasitenkd.*, **72,** 553–556.

Kannan, R., Labotka, R., and Low, P.S. (1988). Isolation and characterization of the hemichrome-stabilized membrane protein aggregates from sickle erythrocytes. *J. Biol. Chem.*, **263,** 13766–13773.

Krugliak, M., Zhang, J., and Ginsburg, H. (2002). Intraerythrocytic *Plasmodium falciparum* utilizes only a fraction of the amino acids derived from the digestion of host cell cytosol for the biosynthesis of its proteins. *Mol. Biochem. Parasitol.*, **119,** 249–256.

Laser, H. and Klein, R. (1979). Haemoglobin S and *P. falciparum* malaria. *Nature*, **280,** 613–614.

Lehmann, H. and Cutbush, M. (1952). Sickle cell trait in southern India. *Br. Med. J.*, **6,** 404–405.

Lell B., May, B, J., Schmidt-Ott, R.J. *et al.* (1999). The role of red blood cell polymorphisms in resistance and susceptibility to malaria. *Clin. Infect. Dis.*, **28,** 794–799.

Liu, S.C., Palek, J., Yi, S.J. *et al.* (1995). Molecular basis of altered red blood cell membrane properties in Southeast Asian ovalocytosis: Role of band 3 protein in band 3 oligomerization and retention by the membrane skeleton. *Blood*, **86,** 349–358.

Livingstone, F.B. (1971). Malaria and human polymorphisms. *Ann. Rev. Genet.*, **5,** 33–64.

Low, P.S., Waugh, S.M., Zinke, K., and Drenckhahn, D. (1985). The role of hemoglobin denaturation and band 3 clustering in red blood cell aging. *Science*, **227,** 531–533.

Lutz, H.U. (1990). Erythrocyte clearance. In: J.R. Harris (ed), *Blood Cells: Subcellular Biochemistry*, Vol. 17, *Erythroid Cells*. Plenum Press, New York, pp. 81–120.

Luzzatto, L., Nwachuku-Jarrett, E.S., and Reddy, S. (1970). Increased sickling of parasitized erythrocytes as mechanism of resistance against malaria in the sickle-cell trait. *Lancet*, **i,** 319–322.

Luzzatto, L., Sodeinde, O., and Martini, G. (1983). Genetic variation in the host and adaptive phenomena in *Plasmodium falciparum* infection. *Ciba Foundation Symp.*, **94,** 159–173.

Luzzatto, L., Usanga, E.A., and Reddy, S. (1969). Glucose-6-phosphate dehydrogenase deficient red cells: Resistance to infection by malarial parasites. *Science*, **164,** 839–842.

Luzzi, G.A., Merry, A.H., Newbold, C.I., Marsh, K., and Pasvol, G. (1991). Protection by alpha-thalassemia against *Plasmodium falciparum* malaria: Modified surface antigen expression rather than impaired growth or cytoadherence. *Immunol. Lett.*, **30,** 233–240.

Luzzi, G.A., Torii, M., Aikawa, M., and Pasvol, G. (1990). Unrestricted growth of *Plasmodium falciparum* in microcytic erythrocytes in iron deficiency and thalassaemia. *Br. J. Haematol.*, **74,** 519–524.

Mackey, J.P. and Vivarelli, F. (1954). Sickle-cell anaemia (Letter). *Br. Med. J.*, **i,** 276.

Mannu, F., Arese, P., Cappellini, M.D. *et al.* (1995). Role of hemichrome binding to erythrocyte membrane in the generation of band 3 alterations in beta thalassemia intermedia erythrocytes. *Blood*, **86,** 2014–2020.

Modiano, D., Luoni, G., Sirima, B.S., *et al.* (2001). Haemoglobin C protects against clinical *Plasmodium falciparum* malaria. *Nature*, **414,** 305–308.

Modiano, G., Morpurgo, G., Terrenato, L. *et al.* (1991). Protection against malaria morbidity: near-fixation of alpha-thalassemia gene in a Nepalese population. *Am. J. Hum. Genet.*, **48,** 390–397.

Mockenhaupt, F.P., Ehrhardt, S., Gellert, S., *et al.* (2004). α^{+}-thalassemia projects from severe malaria in African children. *Blood*, **104**, 2003 - 2006.

Mynt, O., Upston, J.M., Gero, A.M., and O'Sullivan, W.J. (1993). Reduced transport of adenosine in erythrocytes from patients with beta-thalassaemia. *Int. J. Parasitol.*, **23,** 303–307.

O'Donnell, A., Allen, S.J., Mgone, C.S. *et al.* (1998). Red cell morphology and malaria anaemia in children with Southeast-Asian ovalocytosis band 3 in Papua New Guinea. *Br. J. Haematol.*, **101,** 407–412.

Pasvol, G. (1980). The interaction between sickle haemoglobin and the malarial parasite *Plasmodium falciparum*. *Trans. R. Soc. Trop. Med. Hyg.*, **74,** 701–705.

Pasvol, G., Weatherall, D.J., and Wilson, R.J.M. (1978). Cellular mechanism for the protective effect of haemoglobin S against *P. falciparum* malaria. *Nature*, **274,** 701–703.

Pattanapanyasat, K., Yongvanitchit, K., Tongtawe, P. *et al.* (1999). Impairment of *Plasmodium falciparum* growth in thalassemic red blood cells: Further evidence by using biotin labeling and flow cytometry. *Blood*, **93,** 2116–3119.

Roberts, D.J., Harris, T., and Williams, T. (2004). The influence of inherited traits on malaria infection. In: R. Bellamy (ed), *Susceptibility to Infectious Diseases: The Importance of Host Genetics*. Cambridge University Press, Cambridge, UK, pp. 139–184.

Roberts, D.J., Williams, T.N., and Pain, A. (2005). Genetics of resistance to malaria on the Indian subcontinent. In: D.Kumar (ed), *Genetris of Disease on the Indian subcontinent*, in press.

Roth, Jr, E.F., Raventos-Suarez, C., Rinaldi, A., and Nagel, R.L. (1983). Glucose-6-phosphate dehydrogenase deficiency inhibits *in vitro* growth of *Plasmodium falciparum*. *Proc. Natl. Acad. Sci. USA*, **80,** 298–299.

Ruwende, C., Khoo, S.C., Snow, R.W., *et al.* (1995). Natural selection of hemi- and heterozygotes for G6PD deficiency in Africa by resistance to severe malaria. *Nature*, **376,** 246–249.

Sallares, R. (2001). *Malaria and Rome: A History of Malaria in Ancient Italy*. Oxford University Press, Oxford.

Schrier, S.L. (1994). Thalassemia. Pathophysiology of red cell changes. *Annu. Rev. Med.*, **45,** 211–217.

Schwarzer, E., Turrini, F., Ulliers, D. *et al.* (1992). Impairment of macrophage functions after ingestion of *Plasmodium falciparum*-infected erythrocytes or isolated malarial pigment. *J. Exp. Med.*, **176,** 1033–1041.

Scorza, T., Magez, S., Brys, L., and De Baetselier, P. (1999). Hemozoin is a key factor in the induction of malaria-associated immunosuppression. *Parasite Immunol.*, **21,** 545–554.

Senok, A.C., Nelson, E.A.S., Li, K., and Oppenheimer, S.J. (1997). Thalassemia trait, red blood cell age and oxidant stress: Effects on *Plasmodium falciparum* growth and sensitivity to artemisinin. *Trans. R. Soc. Trop. Med. Hyg.*, **91,** 585–589.

Serjeantson, S. Bryson, K., Amato, D., and Babona, D. (1977). Malaria and hereditary ovalocytosis. *Hum. Genet.*, **37,** 161–167.

Siniscalco, M., Bernini, L., Filippi, G. *et al.* (1966). Population genetics of haemoglobin variants, thalassaemia and glucose-6-phosphate dehydrogenase deficiency, with particular reference to the malaria hypothesis. *Bull. WHO*, **34,** 379–93.

Siniscalco, M., Bernini, L., Latte, B., and Motulsky, A.G. (1961). Favism and thalassemia in Sardinia and their relationship to malaria. *Nature*, **190,** 1175–1180.

Spielman, A. and D'Antonio, M. (2001). *Mosquito. A Natural History of Our Most Persistent and Deadly Foe.* Faber and Faber, London.

Tilley, L., Nash, G.B., Jones, G.L., and Sawyer, W.H. (1991). Decreased rotational diffusion of band 3 in Melanesian ovalocytes from Papua, New Guinea. *J. Membr. Biol.*, **121,** 59–66.

Ting, Y.L.T., Naccarato, S., Qualtieri, A. *et al.* (1994). *In vivo* metabolic studies of glucose, ATP and 2,3-DPG in beta-thalassaemia intermedia, heterozygous beta-thalassaemic and normal erythrocytes: ^{13}C and ^{31}P MRS studies. *Br. J. Haematol.*, **88,** 547–554.

Turrini, F., Arese, P., Yuan, J., and Low, P.S. (1991). Clustering of integral membrane proteins of the human erythrocyte membrane stimulates autologous IgG binding, complement deposition, and phagocytosis. *J. Biol. Chem.*, **266,** 23611–23617.

Turrini, F., Ginsburg, H., Bussolino, F. *et al.* (1992). Phagocytosis of *Plasmodium falciparum*-infected human red blood cells by human monocytes: Involvement of immune and nonimmune determinants and dependence on parasite developmental stage. *Blood*, **80,** 801–808.

Udomsangpetch, R., Sueblinvong, T., Pattanapanyasat, K., Dharmkrongat, A., Kittikalayawong, A., and Webster, H.K. (1993). Alteration in cytoadherence and rosetting of *Plasmodium falciparum*-infected thalassemic red blood cells. *Blood*, **82,** 3752–3759.

Urban, B.C. and Roberts, D.J. (2002). Malaria, monocytes, macrophages and myeloid dendritic cells: sticking of infected erythrocytes switches off host cells. *Curr. Opin. Immunol.*, **14,** 458–465.

Weatherall, D.J. (1997). Thalassemia and malaria, revisited. *Ann. Trop. Med. Parasitol.*, **91,** 885–890.

Weatherall, D.J. (1998). Pathophysiology of thalassaemia. *Baillières Clin. Haematol.*, **11,** 127–146.

Willcox, M., Björkman, A., Brohult, J. *et al.* (1983). A case-control study in northern Liberia of *Plasmodium falciparum* malaria in haemoglobin S and beta-thalassaemia trait. *Ann. Trop. Med. Parasitol.*, **77,** 239–246.

Williams, T.N., Maitland, K., Bennett, S. *et al.* (1996). High incidence of malaria in α-thalassaemic children. *Nature*, **383,** 522–525.

Williams, T.N., Wambua, S., Uyoga, S. *et al.* (2005). Both heterozygous and homozygous alpha + thalassemias project against severe and fatal *Plasmodium falciparum* malaria on the coast of Kenya, *Blood*, **106**, 368 – 371

Winograd, E. and Sherman, I.W. (1989a). Characterization of a modified red cell membrane protein expressed on erythrocytes infected with the human malaria parasite *Plasmodium falciparum*: possible role as a cytoadherence mediating protein. *J. Cell. Biol.*, **108,** 23–30.

Winograd, E. and Sherman, I.W. (1989b). Naturally occurring anti-band 3 autoantibodies recognize a high molecular weight protein on the surface of *Plasmodium falciparum* infected erythrocytes. *Biochem. Biophys. Res. Commun.*, **160,** 1357–1363.

Winograd, E., Greenan, J.R.T., and Sherman, I.W. (1987). Expression of senescent antigen on erythrocytes infected with a knobby variant of the human malaria parasite *Plasmodium falciparum*. *Proc. Natl. Acad. Sc. USA*, **84,** 1931–1935.

Yenchitsomanus, P., Summers, K.M., Board, P.G. *et al.* (1986). Alpha-thalassemia in Papua New Guinea. *Hum. Genet.*, **74,** 432–437.

Yuthavong, Y., Butthep, P., Bunyaratvej, A., and Fucharoen, S. (1987). Inhibitory effect of beta^{0}-thalassemia/hamoglobin E erythrocytes on *Plasmodium falciparum* growth in vitro. *Trans. R. Soc. Trop. Med. Hyg.*, **81,** 903–906.

Clinical, Epidemiological, and Genetic Investigations on Thalassemia and Malaria in Italy

Stefano Canali
Gilberto Corbellini

Abstract

The first epidemiological studies on thalassemia distribution in Italy carried out by Ezio Silvestroni and Ida Bianco after World War II confirm the geographic correlation previously reported by different clinicians between the high frequency of this hereditary condition and malarial infection. The phenomenon was studied by the Italian geneticist Giuseppe Montalenti, who stimulated by the hypothesis advanced by J.B.S. Haldane that the frequency could not be due to an abnormal rate of mutation but probably to selective advantage of the thalassemic heterozygotes, began collaborating with Silvestroni and Bianco in order to determine the origin of thalassemia distribution in Italy. In the present article a reconstruction is made of the early studies carried out in Italy on the genetics of thalassemia and a discussion is presented of the first hypothesis concerning the relationship between thalassemia and malaria. An examination is then made of the results and methodological difficulties surrounding the research coordinated by Montalenti at Ferrara, also thanks to funding from the Rockefeller Foundation, as well as those carried out by Ruggero Ceppellini and Marcello Siniscalco in Sardinia.

According to Weatherall and Clegg (Clegg and Weatherall, 1999), and Weatherall (2004b), the question of the relationship between thalassemia and malaria remained uncertain until, thanks to the methods of molecular genetics and the macroepidemiological studies carried out in the Southwest Pacific on alpha-thalassemia, "provided unequivocal evidence for protection." Perhaps

Stefano Canali • University of Cassino, Via Passo Cento Croci, 1-00048 Nettuno (Rome), Italy.
Gilberto Corbellini • Section of History of Medicine, "La Sapienza" University of Rome, Viale dell'Università 34/a 00185, Rome, Italy.

Malaria: Genetic and Evolutionary Aspects, edited by Krishna R. Dronamraju and Paolo Arese, Springer, New York, 2006.

this is true. The fact remains that, historically speaking, the idea that a specific human disease, malaria, might have a selective role and therefore modify the genetic composition of a population by favoring certain genotypes rather than others, was suggested for the first time with reference to the relationship between thalassemia and malaria. The demonstration of the selective action of malaria came, as we know, from Anthony Allison's research on sickle-cell anemia in East Africa (Allison, 1954). Nevertheless, in Italy, the first studies were carried out to determine the possible origin of the high frequency of the thalassemic gene in certain areas of the country, and if the cause could possibly be, as hypothesized by J.B.S. Haldane in 1949 (Haldane 1949a,b), a selective advantage of the heterozygotes due to the concomitant presence of malarial infection. The story of how the "malaria hypothesis," also known as the "Haldane hypothesis," has been pieced together by several authors (Weatherall, 2004a for details, Dronamraju, 2004). Less is known of several premises and consequences of the hypothesis of a causal relationship between malaria and thalassemia, from the standpoint of the epidemiological observations and research carried out in Italy.

1. The Evolution of the Knowledge of Thalassemia Genetics: The Italian Contribution

The exact demonstration of the Mendelian transmission of beta-thalassemia is the result of research carried out between 1940 and 1947 independently in the United States and Italy. Probably for linguistic reasons, literature on the history of thalassemia, with the works of David Weatherall (1980, 2004b, Weatherall and Clegg, 2001) and Maxwell Wintrobe (1980, 1985) in the forefront, has extensively documented the contribution made by US research and displays only a partial knowledge of the research and debate on the genetics of Cooley's disease that had been conducted to a certain extent in Italy ever since the 1920s (for a detailed reconstruction of the italian contribution to the description of the genetic bases of thalassemia see Canali, 2005).

In 1932, when George Whipple suggested the term "thalassemia" (Whipple and Bradford, 1932, 1936), Thomas Cooley noticed that the disease he had clinically defined showed a clear familial incidence (Cooley and Lee, 1932). Later, Heinrich Lehndorf proposed that Cooley's anemia was an inherited disease due to a genetic mutation (Lehndorf, 1936).

The Mendelian inheritance gradually became apparent. Ferdinando Micheli and collaborators (1935), Angelini (1937), and Jean Caminopetros (1937) observed distinctive, albeit minimal, hematological traits in healthy parents. Then Maxwell Wintrobe (1940) and William Dameshek

(1900–1966) (1940) recognized the relationship between these hematological traits and minor forms of thalassemia. Jean Caminopetros (1937) had already assumed that clinically healthy people could transmit the disease as a Mendelian recessive factor.

The Mendelian inheritance of thalassemia was suggested for the first time by Alan Moncrieff and Lionel Ernest Whitby in 1934. Assuming the hereditary condition for thalassemia, Ignazio Gatto (1941, 1942) associated its lethality with homozygosis ("homozygote-disease"), while the heterozygote represented only minor and nonpathological traits ("heterozygote-stigmata") (Gatto, 1942).

Halfway through the 1940s, Italian and US physicians independently demonstrated the mechanism of inheritance. In 1943, Ezio Silvestroni and Ida Bianco, described an inborn and hereditary hematological anomaly in healthy people, which they subsequently called microcythemia (1945a, 1946a). At the same time, Silvestroni and Bianco showed the genetic relationship between thalassemia and microcythemia, studying several people with Cooley's anemia (Silvestroni and Bianco, 1946; 1946–1947). At the end of these investigations, Silvestroni and Bianco documented the Mendelian inheritance of microcythemia as the heterozygotic condition and the homozygotic condition in Cooley's disease (Silvestroni and Bianco, 1947a,b). These results confirmed the evidence obtained in similar studies conducted in the USA by Dameshek (1943) and especially by William Valentine and James Neel at the University of Rochester (Neel and Valentine, 1947; Valentine and Neel, 1944).

Silvestroni and Bianco carried out a series of epidemiological studies all over Italy (Silvestroni and Bianco, 1946–1947, 1947, 1948a,b; Silvestroni *et al.*, 1950), using a specific method to detect the microcythemic trait through the reduction of globular fragility (Silvestroni and Bianco, 1945b). The results of the research, which by 1950 had involved some 50,000 persons, enabled the two to map, for the first time ever, the disease's distribution for a whole country. This distribution revealed a disturbing epidemiological profile, with numerous microcythemic foci scattered all over the country, particularly in the areas of the Po Delta and the islands, Sardinia, and Sicily, where the incidence of carriers exceeded 20% of the population. The map revealed a strong geographic correspondence between the frequency of the thalassemic features and endemic malaria. This singular correlation that had already been observed by clinicians, was now documented by thorough and wide-ranging epidemiological research, thus raising even more clearly the question of maintaining the frequency of a gene that, at the time, doomed homozygotes to death within the first 2 years of life.

2. Early Observations on the Association between Malaria and Thalassemia

The problem of the links between Cooley's anemia and malaria had been debated in Italian medical literature since the 1920s.

Different studies have considered the possibility that malaria was an etiological factor for thalassemia, as it was recognized that a particularly high frequency of cases of Mediterranean anemia existed in a number of malarial zones in families affected by Cooley's anemia (Auricchio, 1928; Careddu, 1929). These coincidences were confirmed by subsequent observations pointing to the presence or the frequency of the disease in malarial zones (Dolce, 1939; Frontali and Rasi, 1939), or the existence of thalassemic foci in Italy (Sicily, Sardinia, Ferrara area, and Puglia) corresponding to endemic malaria zones (Francaviglia, 1939; Auricchio, 1940; Pachioli, 1940). A similar situation was observed in Greece (Choremis and Spiliopoulos, 1936). An even more significant fact was the exact localization of Cooley's anemia in intensely malarial zones, such as Sardinia, in which the frequency of Cooley's disease was very high in the malarial coastal areas and practically absent in the non-malarial internal mountainous zones (Careddu, 1940, 1941). Working at the pediatric clinic of Cagliari University, Cadeddu collected a large number of case histories from all over Sardinia and systematically analyzed all the relevant environmental variables: "in the entire mountain area of the island" – he wrote – "in which malaria occurs sporadically there is an absence of observations except for one observation from Macomer but due to the family community moving there from a malarial zone [. . .]. The towns from which our cases have been drawn almost all have an altitude of less than 100 m above sea level." (Cadeddu, 1942).

The hypotheses regarding the relationship between malaria and thalassemia, put forward by clinical practitioners, who had observed the geographic links between them remained restricted to the medical domain (i.e., at the level of pathogenetic explanation physicians had too little knowledge of genetics and evolutionary biology to allow those working in a clinical environment to elaborate interpretations capable of placing the focus on the links between a population's gene pool and selective pressures).

On the basis of the frequent reports of the malarial plasmodium and the alleged therapeutic effects of quinine in children diagnosed as having thalassemia (in this case it must be recalled that thalassemia was often confused above all in pediatrics with other pathologies having anemia as a symptom), Choremis and Spiliopoulos (1936) postulated that the chronic malaria of the parents triggered a hereditary predisposition of their children's hemopoeietic system. This was then rendered manifested by the superimposition of specific environmental conditions, such as malnutrition. Caminopetros, however, immediately challenged the idea of the direct etiology on the basis of the evidence of the absolute inefficacy of quinine therapy on thalassemic subjects, and followed the completely opposite approach of inoculation of the malaria plasmodium (Caminopetros, 1937).[1]

The hypothesis of a direct etiology had to come to terms with a series of other anomalies pointed out by Gino Frontali and F. Rasi working at the pediatrics clinic of Padua University (Frontali and Rasi, 1939). First, the absence of thalassemia in populations with a high incidence of malaria or conversely the existence of cases of Mediterranean anemia unrelated to any malarial influence, and finally, the inefficacy of antimalarial therapy observed by many clinicians (Vallisneri, 1940). Marino Ortolani, director of the Ferrara Provincial Institute for Childhood pointed out that the hypothesis of direct malarial etiology was based on a body of case histories collected through a systematic diagnostic error that led to a series of obvious cases of malaria being listed as Cooley's anemia and for this reason apparently sensitive to quinine therapy (Ortolani, 1941).

In 1929 Giovanni Careddu suggested an interesting indirect action by malaria on the germ cells of the parents that was capable of leading to an alteration of the hemopoeietic and osteogenetic mesenchyma. Federico Vecchio, at the pediatric clinic of Naples University, in 1947 made explicit reference to germ plasma mutation, referring to Herman Muller's experiments with radiations and the recent (at the time) demonstration that thermal shock seemed capable of producing genetic mutations. It was postulated that gene variation indeed emerged as a consequence of "a direct action of malarial parasitism carried out by the repetition, especially in the prequinine era, of comparatively violent thermal shocks." (Vecchio, 1947, pp. 52–53). In the light of the growing evidence that thalassemia was not exclusively a Mediterranean disease, but one of the most widespread hereditary diseases in the world, Vecchio claimed that the cause should be sought in the exposure "of several ethnic groups to given environmental factors" (Vecchio, 1947, p. 53), rather than in a given genotypic constitution of the populations affected.

---

[1]Caminopetros' malariotherapy of Cooley's anemia was based on an accidentally observed fact. One of his patients, who had accidentally been infected by malaria, displayed a prolonged remission of the erythroblastic reaction. And given that medullary hyperactivity seemed to represent the original pathogenetic stage of Cooley's anemia, Caminopetros decided to induce malaria to slow down the hyperactivity of the bone marrow and thus the erythroblastic reaction and erythroblastosis in the circulation.

Again in 1939, Michele Bufano actually considered the link between malaria and thalassemia a case of inheritance of acquired characters, postulating that malaria produced a series of pathological transformations of the bone marrow capable of being transmitted to the offspring and giving rise to the clinical symptoms of Cooley's anemia. On the other hand, in a long discussion in *Clinica Pediatrica*, Renato Pachioli, a lecturer at Bologna University, suggested viewing Cooley's anemia also as a hereditarily transmissible malarial hemodystrophy originating from a mutation somehow induced by malaria (Pachioli, 1940).

Pachioli claimed that this hypothesis would allow a single interpretative model to be used to explain the apparently hereditary nature of Mediterranean anemia and the singular similarity between several fundamental parts of the pathogenetic processes in malaria and Cooley's anemia, in particular the alterations of erythropoiesis. The explanatory hypothesis was thus constructed by linking together several early speculations on the genetic determinisms of Cooley's anemia with a series of pathological data. On one hand, there was the idea put forward by Heinrich Lehndorff in 1936 that Cooley's anemia is the effect of a genetic mutation, following which the erythroblastic system becomes incapable of producing mature red corpuscles. On the other hand, the (incorrect) observations made above all by Virgilio Chini, seemed to show that malarial infection electively damaged the hemopoietic system and that this was somehow transmitted to the offspring, thus determining bone lesions that could be linked to the pathognomonic lesions of Cooley's anemia. This idea, however, allowed Pachioli to claim that the eradication of malaria would lead to a gradual reduction in the frequency of the thalassemic gene: "the causal relations, although indirect, between malaria and Cooley's anemia [. . .] allow it to be envisaged that the rehabilitation of malaria-infested zones can, in the course of generations, lead to the gradual exhaustion of this morbid hereditary defect" (Pachioli, 1940, p. 422).

Although in a completely nebulous and speculative explanatory framework, Cesare Cocchi put forward the hypothesis that Cooley's anemia, just like favism, represented a defensive process typical of subjects in populations that had been exposed at length to malaria. The idea was that "Cooley's anemia, with its clinical and pathological features, can manifest itself only in subjects prepared by a broad malarial inheritance in the sense that it represents a particular reaction of enhanced defense (instead of increased vulnerability). What I mean is that it is possible to imagine that the same morbid cause (toxic, infectious, or due to deficiency) determines this disease, in that particular way, only in subjects that have malarial forebears, and are thus better protected (we know they are better protected against malaria itself than a subject who is not a descendant of malarial patients), better prepared to defend themselves against all hemolyzing action" (Cocchi, 1941, pp. 286–287).

On the basis of Cocchi's hypothesis, in 1943 Marino Ortolani carried out research "on the immune state regarding malarial infection in subjects, some of whom present the classic Cooley type anemia symptoms and others affected by erythroleukemic myelosis with or without hyperhemolysis" (Ortolani, 1946, p. 38). Ortolani repeatedly tried to graft the "*Plasmodium vivax*" into some patients by inoculating them with blood taken from soldiers from the Greek–Albanian front suffering from malaria. Without stating the number of cases precisely, he reported having failed to infect the subjects involved in the research. Malarial patients lacked the clinical signs of malaria, while the search for parasites in the blood was always negative, even after spleen contraction, as well as the search via medullary and spleen puncture. The only exception was a 10-year-old girl who, after being subjected to a cycle of four inoculations over 4 months of blood that was particularly rich in parasites drawn from four malaria patients, developed malaria 2 years later, displaying clinical symptoms and showing the presence of the benign tertiary plasmodium. The author also reported the case of a baby with Cooley's anemia who died about 2 months after being inoculated with infected blood and who tested negative for the plasmodium even on autopsy via bone marrow examination (ibid, p. 38).

In his comment on these results Ortolani explicitly cited the passage from Cocchi's work mentioned earlier, then claiming that his study seemed to support the hypothesis that Cooley's anemia ought to be considered as a hereditary form of defense against malaria that developed through long exposure of certain populations to the infection (ibid., p. 38). Although unsystematic and ethically dubious, Ortolani's experiment takes on an exceptional historical value as it is preceded by approximately 10 years Allison's work, demonstrating the greater resistance of sickle-cell anemia patients to malarial infection. For reasons that we shall try to understand in the following, Ortolani's experiment was not replicated even in Italy. The only trace of it, we found in the literature was in a 1957 article by Ezio Silvestroni and Ida Bianco on the dissemination of the thalassemic trait (Silvestroni and Bianco, 1957).

The absence of any consistent classification of the pathogenetic mechanisms of thalassemia, as a function of the etiology and pathogenesis of malarial infection, materially stood in the way of any emerging hypothesis that malaria could represent a selective fact, capable of favouring a mutation of the erythropoietic system. Moreover, there was no exact superimposition of the malarial zones on the zones with a high frequency of Cooley's anemia or the microcythemic trait. Reported exceptions were the Rome rural area and the Tuscan Maremma, which for centuries had been intensely malarial but characterized by a low incidence of Mediterranean anemia and microcythemia. In his introduction to the course of medical pathology and clinical methodology at Bari University, Virgilio Chini (1939a), one of the greatest Italian experts in Mediterranean anemia of the time, postulated that the absence of any link between the distribution of malaria and that of Cooley's anemia could be accounted for by biological difference among the types of malaria, distributed through the various different geographic regions. In the same year, Chini reported certain similarities in the radiological examinations of 40 of his malaria patients with those observed in Cooley's anemia (1939b).

In 1941, Franco Toscano reported a geographic correspondence between malaria and Rietti–Greppi–Micheli disease, or thalassemia intermedia, postulating a link with a single etiopathogenesis deriving from ancestral malaria (Toscano, 1941).

Again with reference to the malarial hypothesis, two obvious theoretical anomalies were pointed out: (1) the relatively small number of Cooley's patients also in the more intensely malarial zones and (2) the fact that not all the offspring of chronic malarial patients were affected by Mediterranean anemia. Even though there was no lack of researchers, such as Paradiso (1942), that believed that the fact that Cooley's anemia was found in nonmalarial zones did not rule out the possibility that malaria might be part of the more or less remote ancestry of these patients.

It was also a rather common occurrence to attempt to use ethnic and anthropological factors to account for the distribution of the thalassemic trait and its link to malaria. Ezio Silvestroni, Ida Bianco and Nereo Alfieri, archeologist and director of the Spina Museum in Ferrara, where the two researchers had carried out an intense screening activity, noticed that the regions with a high incidence of thalassemia corresponded to those colonized by the Greeks and that, like certain cities in the Ferrara area or the Po Delta, such as Adria, had intense trading relations with Greece in Roman times (Silvestroni *et al.*, 1952). Although interesting, this hypothesis provided no solution to the problem of the maintenance of high gene frequency. The idea was that the selective pressure of the malaria was subsequently grafted onto the effect of the founder, however, at that stage again raising the problem of the demonstration of the links between malaria and thalassemia. However, it still remained to explain the high frequency of thalassemia in Sardinia, which was inhabited by a population of different ethnic origin, and the high frequency of the trait in the Far East, which was just beginning to be recognized in the research of the time.

3. Collaboration of Silvestroni and Bianco with Montalenti: At Work on Haldane's Hypothesis

In 1947, Silvestroni and Bianco made the first arrangement regarding collaboration with the Institute of Genetics of Naples University, directed by Giuseppe Montalenti.[2] It had been the lack of appreciation and the criticism made by the medical community on the genetic aspects of their work that led the two clinical pathologists to seek the support of a specialist in the field. Montalenti was the most authoritative Italian geneticist. Furthermore, he was already familiar with and had full understanding of the importance of the genetic discoveries made by Silvestroni and Bianco, namely the implications and extraordinary potential for the purpose of new research of the data, collected by these two scientists in the screening campaigns that had begun in the summer of 1943.

Montalenti advised them to investigate the causes of the persistence of the microcythemic foci, identified by Silvestroni and Bianco, in spite of the selection of the thalassemic genes at each generation due to the impossibility of the homzygotes reproducing. He also suggested, studying several well-known Mendelian traits in the populations investigated by the two clinicians, such as sensitivity to phenylthiocarbamide, and also to check any possible link with the microcythemic trait (Silvestroni and Bianco, 1950).

Between the end of July and the beginning of August 1948, just a few months after the beginning of collaboration with the two Roman physicians, who had described the microcythemic condition, Montalenti met John Burdon Sanderson Haldane at Pallanza, on the occasion of a Symposium on the ecological and genetic factors of animal speciation, organized by Adriano Buzzati Traverso in the biophysics section he directed at the Physiopathology Study Centre of the National Research Council. In this forum Haldane presented a report on the links between disease and evolution, which above all hinged on the role of diseases and infectious agents as factors of selection and thus of evolutionary change (Haldane, 1949b). In the discussion that followed, Montalenti asked Haldane if he deemed it hypothetically possible for the disease to have played a role in maintaining the microcythemic gene. Haldane answered

---

[2]Ida Bianco, personal communication to one (S.C.) of the authors. (Also see Archivio Giuseppe Montalenti)

affirmatively, adding that the advantage of the microcythemic heterozygote could also be derived from an enhanced capacity for iron absorption in populations with diets deficient in this element.

Haldane's article "*Disease and Evolution*" is the one most frequently cited as that in which the malaria hypothesis, also known as the Haldane hypothesis, was first formulated even though in that context it was not Haldane but Montalenti who first suggested the link between thalassemia and malaria. In fact, several months earlier, the English geneticist had already put forward the hypothesis at the International Genetics Congress held on 7–14 July 1948 at Stockholm (Haldane 1949a). When discussing the problem of the mutation rate in man with reference to Neel and Valentine's studies on thalassemia, namely the hypothesis advanced by the two US geneticists that, in order to maintain such a high frequency, the mutation rate would have to be 1:2500 and that this rate might have an ethnic basis (Neel and Valentine, 1947), Haldane considered the rate to be too high, proposing that the erythrocytes of thalassemics were resistant to malaria parasites and the disease's high frequency was the result of a selective heterozygote advantage.

The Haldane hypothesis emerged as an obvious alternative once it could be ruled out that polymorphism was due to a "special" mutation rate. It is not clear when the idea of this hypothesis came to Haldane. On the basis of the reading of the article *Disease and Evolution*, published by the Italian review *La Ricerca Scientifica*, Allison gave credit to Montalenti for suggesting the malaria hypothesis to Haldane (Allison, 2004). According to Weatherall, the hypothesis was originally conceived of at Stockholm, during the World Genetics Congress held several weeks before that of Pallanza. In fact, Weatherall's thesis is not documented, in the sense that it must in any case be demonstrated that the address published in the proceedings of the Stockholm Congress corresponds to that delivered publicly at the Congress. Some doubt as to whether Haldane got the idea before Pallanza seems to emerge from the fact that in the article *Disease and Evolution* he does not propose the example. Why? If he had already presented it at Stockholm, where also Montalenti was present, why did he not talk about it again in Italy? And was it Montalenti who reminded him of it? The fact is that, for the molecular genetics historian and philosopher Sahotra Sarkar there is "archival evidence" that Haldane developed the hypothesis before going to Stockholm[3].

Could it not be that, after his discussion with Montalenti, Haldane modified the written text of his address to the Stockholm Congress,

---

[3]Sahotra Sarkar, personal communication to one of the authors (G.C.), also see Weatherall's contribution to Dronamraju (2004).

including the hypothesis of heterozygote advantage vis-à-vis malarial infection? The question can be resolved only if documentary evidence is available concerning what Haldane actually said in his report to the Stockholm Congress.

The first results produced by the collaboration between Silvestroni and Bianco and Montalenti and his Naples Genetics Institute were published on 29 April 1950 in *Nature*. In the light of the epidemiological data the question arose of the high incidence of the thalassemic gene (indicated as *M* in the article, as opposed to *m,* the symbol of the normal hemoglobin gene) despite its constant elimination with the death of the homozygotes prior to the reproductive age. The authors advanced four hypotheses: "(1) the gene might have, in the heterozygous condition, a positive selective value; (2) the mutation frequency from *m* to *M* might be such as to balance the loss; (3) mating may not occur at random; and (4) the fertility of some genotypes may be higher."

These actually rejected the possibility of the heterozygotes enjoying any selective advantage, as they seemed to contradict the evidence of Silvestroni and Bianco with reference to the health conditions and average age at the death of microcythemics, the former being far worse and the latter much lower than the average for subjects that are not carriers of the thalassemic trait. The second possibility seemed equally unlikely in view of the excessively high frequency of mutation required to maintain the incidence observed in the thalassemic gene. And in any case, even if the hypothesis of such a high mutation frequency was ultimately accepted, it would become extremely difficult to explain why it was so high in certain geographic areas and not in others. Silvestroni and Bianco's data also clashed with the third hypothesis, that of nonrandom mating (i.e., the possible tendency of microcythemics to mate among themselves). The same evidence seemed instead to support the fourth hypothesis, that of the heterozygote genotype, leading to higher fertility since the heterozygous couples had an average number of children that was appreciably higher than that of couples with no thalassemic traits and those of homozygous couples.

The results, the hypotheses, and the conclusions regarding the causes of the balancing of selection of the thalassemic gene were again proposed 2 years later in a broader-based report published in *Eugenics* (Bianco *et al.*, 1952). Numerous theoretical difficulties related to the level of knowledge available at the time stood in the way of any consideration of Haldane's hypothesis. One first problem was the large phenotypic variability of carriers of the thalassemic traits. Furthermore, nothing was known about the complex heterogeneous nature of genetic and molecular phenomena underlying thalassemic phenotypes. Thinking and debate centered around a microcythemic gene, around a single defect in a specific gene, based on the idea of an extremely simple gene. For example nothing at all was known about the existence of alpha-thalassemia, which was instead typical of the Sardinian populations being studied. The idea of a disease that is today classified today as part of a set of different hematological disorders and associated with almost 300 different mutations, could not even be remotely contemplated.

And then up to what extent was gene expression influenced by other genes of the environment? In 1941, Haldane demonstrated that it was possible to determine the origin of the variable expression of a gene by measuring the intensity of character correlation between parents and offspring or among siblings. In Haldane's view, multiple allelism occurred when a correlation coefficient close to one was observed. If instead the intermediate value approached 0.5, the likelihood of gene modifiers became quite possible. With lower correlation coefficients the phenotypic variability must have depended on nongenetic factors. The available information in this connection was, at the time, largely inadequate precisely because of the difficulty of defining and measuring the phenotypic traits to be linked together. In the case of globular resistance, the only trait that could be measured with sufficient precision, a correlation of about 0.30 emerged, at the limit of a significant scale of relations. The investigations thus seemed to point to the action of modifier genes in the variability of phenotypic expression, as anticipated by Silvestroni in 1949 (1949).

In 1953, at Bellagio, on the occasion of the IX International Genetics Congress, when presenting a long report on microcythemia genetics based on Silvestroni and Bianco's research, Montalenti continued to express doubts regarding the idea of a possible heterozygote advantage. The first reason was the partial superimposability in Italy of the epidemiological map of microcythemia on that of malaria. This is a topic that, as we have already seen, had been deemed to be central by clinicians in order to refute the hypothesis of possible links between malaria and Cooley's anemia. Another argument against the hypothesis was the total absence of any causal explanation of the way in which the microcythemic gene could increase resistance to malarial infection. Understanding of the possible heterozygote advantage in this sense could be provided at the level of macroscopic functions. From this point of view it was practically impossible to identify any characteristic traits of the heterozygote, such as would lead to a selective advantage. Furthermore, this opinion was shared at the time by Neel who wrote that: "in our own experience individuals with thalassemia minor have averaged 2 g of hemoglobin less than normal persons. While there is undoubtedly a large margin of safety in normal hematological physiology, it is difficult to see how such a departure from the norm can per se be of adaptive value to the organism. The possibility remains that the hematological trait is linked to some yet unrecognized characteristic of distinct value" (Neel, 1951).

Above and beyond the theoretical aspects of the problem, the group coordinated by Montalenti addressed the issue of microcythemic gene maintenance by seeking a possible alternative survival of microcythemics vs noncarriers. A preliminary analysis was carried out at the end of 1952, on the data collected during the 1951 and 1952 campaigns in the Ferrara area, with a total of about 9000 subjects being tested. As Montalenti himself (1953) acknowledged, there were gaps in the investigation. The older age group seemed underrepresented, and there were few data on subjects under the age of 30. Furthermore, data were collected on married couples with both spouses living, thus making it possible that any selective advantage of one of the two genotypes would be overshadowed. Although controversial, the data nevertheless seem to indicate a greater fitness of carriers of the microcythemic gene compared with normal subjects and thus suggested the need to develop research in that direction.

In 1953, a more thorough examination of couples with genotypes *mm*×*mm*, *mm*×*Mm*, *Mm*×*Mm*, carried out on about 2800 families, did not show up any significant difference in fertility. This seemed to refute the initial hypothesis that the maintenance of the microcythemic gene was indeed due to a greater fertility of couples carrying the thalassemic trait. With the data observed in the same families studied between 1951 and 1952, a complex analysis of the fitness of the different genotypes was also attempted. The idea was to be able to calculate the theoretical fitness required to maintain in equilibrium a given gene frequency in a population of which the gene elimination rate was known. Also in this case, the computations, in any case incorrect in a preliminary work published in *Nature* in 1954 (Silvestroni et al., 1954), produced controversial results, which made it impossible to adequately demonstrate the advantage of heterozygosis, but nevertheless suggested pursuing research in that direction.

A statistical survey carried out to study microcythemia distribution as a function of age seemed indirectly to lean in favor of an advantage of the heterozygote vs malarial infection, as the thalassemic trait was more frequent on average among individuals over the age of 40. It must be pointed out that, in 1946–1947, a 5-year plan had been launched in Italy to eradicate malaria by means of DDT spraying, which had practically eliminated *P. falciparum* starting from 1951. As a result, the selection of malaria to maintain the polymorphism of the microcythemic trait disappeared and the population were tending towards a different equilibrium (Research project, AM B125).

4. Research of Carcassi *et al.*

During the XX Symposium held at Cold Spring Harbor on the topic "Population Genetics: The Nature and Causes of Genetic Variability in Population", Anthony C. Allison summarized the data on polymorphism in man on the basis of his discovery, reported in the previous year, that polymorphism balanced by a sickle-cell trait is due to a selective advantage of the heterozygote vs the serious malaria form (*P. falciparum*). The Italian geneticist Ruggero Ceppellini took part in the discussion, summarizing the genetic studies on thalassemia carried out in Italy and reporting for the first time the results of research carried out in 1954 in Sardinia (Ceppellini, 1955). The experiment carried out by Cercassi *et al.* was extremely elegant and was based on the studies that had allowed Allison to demonstrate the role of *P. falciparum* as a factor responsible for the high frequency of the sickle cell trait in the zones he had studied in East Africa (Allison, 2004). In practice, the Italian researchers made comparisons among the inhabitants of four villages in the province of Nuoro who had been relatively isolated from the reproductive point of view. Two of these villages, Orosei and Saltelli, lay at the bottom of a valley that had been intensely malarial until 1947, with one of the highest rates of *P. falciparum* infection in Europe. In these villages, the percentage of microcythemics was around 20% (18.8 and 21.3, respectively). The other two villages, Desulo and Tonara, were situated quite close as the crow flies (about 50 km), although at an altitude of 1000 m and had had only rare cases of malaria. The frequency of microcythemics among the inhabitants of these two villages was about 5% (3.75 and 4.67%, respectively.)

It is significant that these researchers were the first to grasp that the observation made as early as 1940, concerning the coincidence of the absence of malaria and thalassemia at higher altitudes, if suitable genetically isolated communities were selected, represented a useful context in which it was possible to rule out environmental factors other than malaria. Having eliminated on the basis of a thorough study of the frequency of blood groups, the hypothesis of ethnic heterogeneity as a possible explanation of the different frequency of the microcythemic trait, Cercassi *et al.* claimed that the only hypothesis displaying favorable empirical elements is that malarial infection exerted a selective pressure to the advantage of microcythemics (heterozygotes). "The fundamental similarity, said Ceppellini in 1955, of dietary, social, and economic conditions points toward malaria as the most important environmental difference between the two groups. Naturally, the first data are only of limited value, and more careful and extensive investigations must be carried out before malaria can be accepted as the environmental agent responsible for the high value of the mutant gene established in the two valley villages" (Ceppellini, 1955, p. 253).

The results of their research were presented in full in a 1957 publication. The authors claimed that the gene frequencies of the thalassemic allele in the plain land villages could be maintained at the high values observed "only if the heterozygote, in the given environment represented by the two countries in question, is advantaged with respect to the normal subject by a selective coefficient of around 12.5%" (Cercassi et al., 1957, p. 211). They emphasized that *Plasmodium* infection was "the element of an environmental nature that more conspicuously diversified the two zones" (plain land and mountain), and was "instead uniform for couples in villages inside the zones" (ib. p. 212). In a more expert fashion they dwelled upon excluding the hypothesis of an ethnic heterogeneity through an analysis of blood group frequency, which indeed they succeeded in demonstrating.

The research carried out by Cercassi *et al.* defined the experimental framework within which the Montalenti group was to work during the late 1950s, also thanks to funding from the Rockefeller Foundation.

5. Studies on Microcythemia Genetics in Italy Funded by the Rockefeller Foundation

In 1954, Giuseppe Montalenti requested funding for a multiyear research program,[4] stressing the privileged condition in which Italy now found itself in the field of thalassemia research owing to the peculiar geographic distribution of this disorder and the quantity of data already collected, as well as the acquisitions of Silvestroni and Bianco (Montalenti, Preliminary draft for a research plan on the genetics of mycrocythemia, AM, B125).

For Montalenti, the first problem to solve was "the cause of the high frequency of the gene in some regions; fluctuations of frequency in time should, if possible, be ascertained. Cause of the maintenance of high frequency in spite of continuous elimination of the gene (lethal in

[4]Already in 1951 Montalenti had obtained a grant from the Rockefeller Foundation to study microcythemia genetics.

homozygous condition) should be investigated. A selective advantage to peculiar environmental conditions (malaria? food quality and/or quantity?) or in regard to other hereditary traits (especially, blood groups and anthropological traits) should be investigated as fully as possible. This was the necessary premise for any action toward the control of a hereditary disease due to a single gene, such as Cooley's disease (thalassemia major)". (Montalenti, Preliminary draft for a research plan on the genetics of mycrocythemia, AM, B125).

The importance of this research and the results did get the notice of the Rockefeller Foundation board. Through R.R. Struthers, director of the European bureau in Paris, he actually suggested to Montalenti that he should increase the research funds available to Silvestroni and Bianco and to fund their studies directly, and no longer through Naples University.[i] Furthermore, Struthers suggested boosting the development of "the eugenic aspect of the microcythemic problem, the establishment of official registers of persons carrying this gene, marriage counseling in some form".[ii] This was precisely what also Silvestroni and Bianco were most interested in, and had already started doing.

In May 1954, Montalenti succeeded in obtaining 5-year funding for a total of 15 million lire (the equivalent of about euros 195,500 in July 2004) to be applied largely to microcythemia studies.[iii] Contrary to Struthers' suggestions, Montalenti decided to use the grant almost exclusively to fund basic research, ignoring the clinical and eugenic aspects.[iv] Moreover, he managed the 5-year fund directly, thus ruling out any possibility of setting up a direct funding channel between the Rockefeller Foundation and the Silvestroni–Bianco group.

The research actually got under way in 1956, the "essential and conditional" point was that "the research funded by the Rockefeller Foundation was supposed to provide the scientific basis for a possible check of the frequency of the "*M*" gene in a limited population. Therefore the entire investigation must be aimed at extending knowledge of micro-

---

[i]Communication of 22 January 1954 from the director of the European Office, R.R. Struthers, to Montalenti. B125, Fondo Montalenti, Bib. Storia della Medicina, Università degli Studi di Roma "La Sapienza."

[ii]Communication of 22 January 1954 from the director of the European Office, R.R. Struthers, to Montalenti. B125, Fondo Montalenti, Bib. Storia della Medicina, Università degli Studi di Roma "La Sapienza."

[iii]Communication of 24 May 1954 from the secretary of the Rockfeller Foundation, Flora M. Rhind to Chancellor and director of the Genetics Institute of Naples University. B125, Archivio Montalenti, Bib. Storia della Medicina, Università degli Studi di Roma "La Sapienza."

[iv]In this connection see research programmes and annual reports on the work performed in B125, Fondo Montalenti, Bib. Storia della Medicina, Università degli Studi di Roma "La Sapienza."

cythemia genetics, of variations in gene frequency in the populations and on the possible influence of the environment on gene manifestation" (Montalenti, Programma di ricerche sulla Microcythemia 1956–1959, AM B125).

The preliminary plan was subdivided into a series of preparatory investigations and into the study of the possible selective advantage of heterozygotes. The former would determine the demographic characteristics of the population in one or more "control" centers, measure the essential anthropological data, study population movement, estimate the coefficient of consanguinity, determine the degree of completeness of the data collected on the distribution of the '*M*' gene, determine the number of Cooley's anemia patients in the population, and examine its agreement with the theoretical premises and lastly perfect diagnosis by means of biochemical examination.

On the basis of Montalenti's general instructions, Silvestroni and Bianco drew up a preliminary research program involving a series of preparatory investigations of human hemoglobins for the purpose of a more accurate biochemical, as well as hematological, identification of true microcythemics as opposed to possible carriers of other abnormal hemoglobins, and to be able to distinguish heterozygous from homozygous microcythemics (Research program on microcythemia and collaboration plan for the first year of Rockefeller funding, AM, B125).

The program also entailed a summer research campaign, aimed at making an examination as complete as possible for the entire population of a new village in the Ferrara, area having a high percentage of microcythemics. The data thus obtained would shed new light on the problem of random mating and on computing the coefficient of consanguinity. The campaign thus involved an anamnestic investigation of the families and consultation of the municipal records in order obtain fresh data for the study of fertility in marriages, death, and birth rates at the various ages, the differential morbidity between normal subjects and microcythemics, the causes of death and to determine the real number of deaths due to Cooley's anemia in the population examined (Research program on microcythemia and plan of collaboration for the first year of Rockefeller funding, AM, B125).

Finally, Silvestroni and Bianco developed the blood smear technique for studying the intensity of the morphological manifestations of the microcythemic gene, to be carried out in accordance with the following three directives:

(1) Study of the intensity of the hereditary line characters (between parents and offspring; in three or more successive generations; among siblings)

(2) Study of the intensity of the characters as a function of age and gender
(3) Study of the intensity of the characters as a function of economic and environmental conditions

In the first 6 months of 1956 Silvestroni and Bianco carried out the preparatory investigations of human hemoglobins. During the summer of the same year they then carried out a research campaign at Berra, a municipality in the province of Ferrara with a population of 4150 (Silvestroni and Bianco, 1957). Silvestroni and Bianco examined the whole population, at the same time carrying out a study of the municipal records, a series of blood tests on samples taken from microcythemic subjects, the anthropological examination of the test subjects, investigation of the links between the hematological and anthropological characters between parents and offspring and among siblings. Owing to their completeness and the extension of the research to a circumscribed population, the data obtained represent study material of exception genetic–statistic and hematological interest, above when it is considered that the two researchers had already "typed" the population of other municipalities in the province including Pomposa, Codigoro, Caprile and other towns for a total of about 50,000 subjects examined.

The issue of determining the existence of a possible heterozygote advantage becomes central in the program of investigations funded by the Rockefeller Foundation. Among the documents conserved in the Montalenti records in the envelope of the so-called "Microcythemia" fund, the only research project drafted separately and not included in the more general research programs conserved in the Microcythemia fund refers precisely to questions concerning Haldane's malaria hypothesis (Research project, AM B125).

The research project specifically concerning Haldane's hypothesis was divided into three separate investigations:

(1) Ascertainment of heterozygote fitness vs the fitness of normal homozygotes
(2) The relationship between microcythemic genotype and death due to malaria
(3) Verification whether death due to malaria was sufficient to guarantee the observed frequencies of microcythemia

These were complex investigations involving different dimensions of the phenomenon and that require sifting through variables that were hard to quantify. In order to determine the fitness of heterozygotes vs normal homozygotes, the research project for example required the construction of mortality rate tables for homozygotes and heterozygotes and thus a comparison of the differential survival of the two genotypes. This made it necessary to determine the fate of individuals "typed" during the various screening campaigns carried out by Silvestroni and Bianco and ultimately discover the cause of death or, if they were married with children and thus if the children were carriers or not. This was an extremely complex task, considering the strong migratory flows from the towns that had been subjected to "hematological census" and the difficulty encountered by the municipalities in recording, reconstructing, and reporting the histories and movements of their citizens. Since the total set of these data referring to at least 10,000 individuals over a period of 5 years was available, it was theoretically possible to ascertain any advantage or disadvantage of the heterozygote also for causes other than malaria.

An even more complex task was the determination between microcythemic genotype and death due to malaria. In this case the research project involved the complete registration, from personal data to the genotype, of all typed individuals above the age of 50. Furthermore, for each of these individuals, the registry records were searched for the presence of siblings who had died of malaria, if possible, going back as far as the prequinine area. This task proved particularly difficult and often impossible owing to wartime damage to the municipal registers (Research project, AM B125).

Similar difficulties accompanied the verification of whether death due to malaria was sufficient to guarantee the observed frequencies of microcythemia. In this case, the research project entailed the collection of the number of deaths due to malaria and of those due to all other causes, going back in time even as far as the nineteenth century and thus to the prequinine era. The data collected in this way were then classified in a double entry table, by year of death and age. The classification by age had to be done painstakingly and, up to 5 years of age, with a class for each year. The aim of this investigation was to relate the frequency of microcythemia in the older age groups to the progress and death rate due to malaria before the introduction and spread of quinine. Indeed, theoretically, microcythemic frequency should be a function of the death rate due to malaria before the advent of quinine and at the same time the death rate trend due to malaria should be related to the effect of age on the frequency of microcythemia in the present population (Research Project, AM B125).

In spite of several inevitable gaps due to the complexity of the research, the processing of the data obtained through the complete observation of several Ferrara centers produced results that supported the hypothesis of malaria as the environmental cause of the high frequency of the gene that was lethal in the homozygous state.

6. The Results of Siniscalco's Genetic Studies on the Distribution of Thalassemia, G-6-PD Deficit, and Malaria

The experience gained in the Ferrara area was of fundamental importance in the development of subsequent investigations. However, the experience of Carcassi *et al.* had shown that the natural laboratory for studying the problem was Sardinia. A population having a peculiar genetic inheritance, the result of geographic isolation, with peculiar cultural characteristics and settlement habits, such as the existence of small villages' isolated for geographic and cultural reasons with populations having a high degree of kinship.

By the end of 1956, Marcello Siniscalco, who was planning and implementing research, aimed at studying the interactions at genetic and population levels between thalassemia, favism (G-6-PD deficit), and malaria. Carrying out a data collection campaign in 19 Sardinian villages (Siniscalco et al., 1961) and later in 52 (Siniscalco et al., 1966), he showed that in the various geographic districts the high incidence of thalassemia and favism correlated positively with previous malarial morbidity, and negatively with the altitude at which the villages were situated. This morbidity was recognized on the basis of epidemiological investigations carried out in the 1930s when malaria was raging on the island of Sardinia. For example, vs frequencies of 3–4% of these hereditary conditions observed in the mountain villages of Gennargentu, such as Fonni, Desulo, Tonara, and Lanusei, frequencies as high as 15–20% were found in Baronia and over 30% in Campidano (Siniscalco et al. 1961; Siniscalco et al., 1966).

The analysis and interpretation of the data collected by Siniscalco were criticized by the anthropologist Peter Brown. In particular, Brown (1981) challenged the assumption underlying Siniscalco's study, namely the negative correlation between thalassemia, G-6-PD deficit, and village altitude, assuming that altitude was sufficient as a substitute measure of the morbidity due to malaria. Working on data obtained directly from the investigations of the hygienist Claudio Fermi between 1933 and 1937, Brown showed that morbidity rates cannot be considered a direct function of altitude and that on the basis of the correlations with actual morbidity no statistically significant relations emerged between the distribution of thalassemia, G-6-PD deficit, and malaria in Sardinia.

Brown correctly demonstrated that it is not the altitude directly, but the ecology of the vector, which can account for the correlations with altitude in so far as the temperature and the presence of stagnant water affect the spread and behavior of *Anopheles labranchiae*. On the grounds of the archaeology and ancient history of Sardinia, Brown postulated that the mutations responsible for thalassemia and favism were introduced during the Carthaginian conquest in the fifth century B.C. and the deforestation of large areas of the island to grow wheat created a favorable habitat for the vector of the malaria parasite. This in turn was responsible for maintaining a high gene frequency. In essence, Brown's thesis is that malaria is not the only explanation for the distribution of thalassemia and favism, but that there was an external gene flow on which the natural selection due to malaria was grafted.

7. The Malaria Hypothesis and the Consequences of Eradicating the Plasmodia and of Thalassemia Prevention

As, according to Haldane's hypothesis, malaria was the selective factor underlying the increased frequency in a population of the thalassemic or microcythemic trait, it was obviously wondered what the consequences might be on the spread of the genetic variant of the eradication of malaria from Italian territory. "If the malaria hypothesis as a selective ecological factor in favor of thalassemia is correct, now that malaria has been completely eradicated as a cause of death in Sardinia, it would be expected that the gene frequency would decrease rapidly. Starting from a frequency of 7% in the present population the frequency should be halved in the course of just seven or eight generations". This is what Latte wrote about the situation in Sardinia in 1968. He added that the elimination of the gene would be speeded up considerably if thalassemic couples, conscious of the risk of giving birth to offspring affected by Cooley's anemia, voluntarily abstained from having children. However, it was not sufficient, as Latte pointed out, not to have mating between thalassemics to avoid the birth of children suffering from Cooley's anemia. "This is of course a eugenic measure for the individual and for the individual family, but not for the species, as in this case the frequency of the gene would remain unaltered in the course of the generations, or else might even increase if the biological fitness of heterozygous carriers were still high". The "biological fitness of the heterozygote genotype" should have been estimated under the existing conditions. Ultimately, a "perhaps less repressive eugenic measure could be to favor the emigration of individual affected family nuclei from microcythemic areas to nonmicrocythemic areas, and vice versa the immigration of normal family nuclei to thalassemic areas." These movements were possible inside Sardinia itself where zones "with a high frequency of thalassemia and villages with very low frequencies" had been clearly identified. (Latte, 1968, pp. 410–411).

The problem of increased microcythemia frequency due to preventive action was instead discussed by Ida Bianco's group. It had been postulated, on the basis of certain computations to estimate the impact of a malarial focus on microcythemic gene frequencies, such as to produce particularly high frequencies even in a comparatively short time (25–30 generations), that the flooding of the Pomposa area after the Po river broke its banks in 1150 and caused the rapid formation of many stagnant water pools might had led to selective pressure to increase the frequency of the microcythemic gene, already present through importation or the result of autochthonous mutation. The very person who had put forward this hypothesis, Robin Bannerman (1961), observed that microcythemia in the world was subject to recurrent fluctuations of environmental factors that, over a period of several centuries, could lead to the appearance or regression of microcythemic foci. In this sense, while "the recent land reclamation and the consequent disappearance of endemic malaria must already have broken the preceding gene equilibrium in the populations and also produced a trend toward the reduction in the gene frequencies of microcythemia; consequently, the introduction of premarriage prophylaxis in a system like this that is no longer in equilibrium but displays a descending trend, will fail to produce new increases but at best will halt the decrease in gene frequencies at the present levels" (Bianco et al., 1976, p. 529).

8. Conclusions

After 1949, the year in which J.B.S. Haldane put forward the hypothesis that the distribution of thalassemia in Italy was the result of a thalassemic heterozygote advantage, the Italian geneticists Giuseppe Montalenti, Ruggero Cappellini, and Marcello Siniscalco carried out a series of studies to verify this hypothesis. The first studies were carried out in collaboration with the clinicians Ezio Silvestroni and Ida Bianco who had been the first to define the genetic bases and the distribution of thalassemia in Italia, while subsequent, more targeted, research was carried out in Sardinia.

The results could not be considered as definitive proof that the thalassemic trait provides protection from malaria above all because of the epidemiologically and genetically complex signs displayed by thalassemia. Nevertheless, by means of this research a huge quantity of data was collected and analyzed, and the cognitive and politico-cultural foundations laid for an antithalassemia campaign to be launched (Canali and Corbellini, 2003).

Acknowledgements

The authors wish to thank Ida Bianco for her help and participation in the ongoing research into the history of thalassemia in Italy, and in particular for having made available her invaluable records.

Stefano Canali also wishes to thank the Wellcome Trust Centre for the History of Medicine at University College London for the hospitality accorded to him in the Summer of 2002 and in 2003.

References

Allison, A.C. (1954). Malaria and sickle-cell anemia, *BMJ*, **1**, 290–294.

Allison, A.C. (2004). Two lessons from the interface of genetics and medicine. *Genetics*, **166,** 1591–1599.

Archivio Giuseppe Montalenti, Biblioteca di Storia della Medicina, Università degli Studi di Roma "La Sapienza" Dipartimento di Medicina Sperimentale e Patologia, Sezione di Storia della Medicina.

Auricchio, L. (1928). Su alcune sindromi di anemia con splenomeglaia a carattere familiare nell'infanzia. *La Pediatria*, **36,** 1023–1040.

Auricchio, L. (1940). Sindromi eritroblastiche nell'infanzia. *Atti del XVII Congresso Italiano di Pediatria*, Roma, 27–29 Settembre 1939, pp. 103–120.

Bannerman, R.M. (1961) *Thalassemia: A survey of some aspects*. Grune & Stratton, New York.

Bianco, I., Graziani, B., and Marini, G.P. (1976). La profilassi delle malattie microcitemiche, in *Atti del V Congresso sulle microcitemie*, Cosenza 4–6 Ottobre 1973, Istituto Italiano di Medicina Sociale, Roma.

Bianco, I., Montalenti, G., Silvestroni, E., and Siniscalco, M. (1952). Further data on genetics of microcythaemia or thalassaemia minor and Cooley's disease or thalassaemia Major. *Eugenics*, **16**(4), 299–315.

Brown, P.J. (1981). New considerations on the distribution of Malaria, thalassemia and glucose-6-phosphate dehydrogenase deficiency in Sardiana. *Hum. Biol.*, **53**(3), 367–382.

Bufano, M. (1939). Comunicazione Relaz. Di Guglielmo. *Haematologica* **17,** 43.

Cadeddu, G. (1942). Fattori ambientali e costituzionali nella malattia di Cooley. *Medicina infantile*, **13**(3), 49–54.

Caminopetros J (1937) L'anémie érythroblastique des peuples de la méditerranée orientale, *Monographies de l'Académie d'Athénes*, 6, 3: 81–143

Canali, S. (2005). From splenic anemia in infancy to Microcythemia. The italian contribution to the description of the genetic bases of thalassemia. *Medicina nei secoli,* **17**(1), 161–179.

Canali S. and Corbellini, G. (2003). Lessons from anti-thalassemia campaigns in Italy, before prenatal diagnosis, *Medicina nei secoli,* **14** (3), 739–771.

Careddu, G. (1929). Anemia splenica infantile e terapia attinica. *Rivista di Clinica Pediatrica*, **20**(7), 1–25.

Careddu, G. (1940). Osservazioni sulla anemia di Cooley in gemelli. *Studi Sassaresi. Sezione II: Scienze Mediche e Naturali*, **18,** 3–8.

Careddu, G. (1941). A proposito del fattore razziale nell'anemia splenica di Cooley. *Atti del XVII Congresso Italiano di Pediatria*, Napoli 20–25 Maggio 1940, parte II, Arti Grafiche Vasesiane G.B. Capelli, Varallo Sesia, pp. 506–507.

Ceppellini, R. (1955). Discussion. Cold Spring Harbor Symposia on Quantitative Biology, **20,** 252–255.

Cercassi, U, Ceppellini, R., and Pitzus, F. (1957). Frequenze della talassemia in quattro popolazioni sarde e suoi rapporti con la distribuzione dei gruppi sanguigni e della malaria. *Bollettino dell'istituto istituto sieroterapico milanese*, **36,** 206–218.

Chini, V. (1939a). Orientamenti moderni di clinica ematologica e loro rapporti con l' "Anemia mediterranea" nei suoi riflessi storici e sociali". *Policlino*, (Sez. Pratica), **46**.

Chini, V. (1939b). Su alcuni rapporti tra infezione malarica e sindromi tipo Cooley. *Haematologica*, **20,** 1–9.

Choremis, C. and Spiliopoulos, J. (1936). Ueber die Cooleysche Anämie. *Kinderärztliche Praxis*, **12**.

Choremis, C. and Spiliopoulos, J. (1937). Ueber die Aetiologie und Therapie der Cooleyschen Anaemie. *Jahrb. f. Kinderheilk*, **148,** 317.

Clegg J.B. and Weatherall D.J. (1999). Thalassemia and malaria: new insights into an old problem. *Proc. Assoc. Am. Physicians*., **111**(4), 278–282.

Cocchi, C. (1941). L'anemia mediterranea di Cooley. Nuovi casi osservati in Sardegna: tentativi di terapia. Ipotesi etiologiche e patogenetiche. *Rivista di Clinica Pediatrica*, **39**(5), 257–287.

Cooley TB, Lee P (1932) «Erythroblastic anemia» *American Journal of Disease of Children*, 43: 705–708.

Dameshek W (1940) "Target cell" anemia. Anerythroblastic type of Cooley's erithroblastic anemia. *American Journal of Medical Sciences*, 200: 445–454.

Dameshek W (1943). Familial mediterranean target-oval cell syndromes. *American Journal of Medical Sciences*, **205,** 643–660.

Dolce, N. (1939). Distribuzione geografica dell'anemia splenica (tipo Cooley) in Italia. Soc. Coop. Tip. Padova.

Dronamraju, K.R. (ed) (2004). *Infectious Disease and Host-Pathogen Evolution*. Cambridge University Press, Cambridge, UK.

Francaviglia, A. (1939). Ricerche sul morbo di Cooley. I: Studio su alcuni casi di morbo di Cooley osservati in Puglia. *Archivio per le scienze mediche*, **68**(4), 395–408.

Frontali, F. and Rasi, F. (1939). L'eritroblastosi e l'emolisi nella malattia di Cooley e di Di Guglielmo. *Archivio Italiano di Pediatria e di Puericultura*, **7,** 259–345.

Gatto I (1941) Ricerche sui familiari dei bambini affetti da malattia di Cooley. Contributo alla comprensione della ereditarietà del male. *Sezione Siciliana della Società Italiana di Pediatria*, 12, 12. *Rivista di Clinica Pediatrica*, 14, 5: 225–252.

Gatto I (1942) Ricerche sui familiari dei bambini affetti da malattia di Cooley. *Archivio Italiano di Pediatria e Puericultura*, 9: 128–168.

Haldane, J.B.S. (1941). The relative importance of principal and modifying genes in determining some human diseases. *J. Genet*., **41,** 149–157.

Haldane JBS (1949a) The rate of mutation of human genes. Proceedings VIIIth International Congress of Genetics, 1948, Stockolm, in *Hereditas*, 1949, Suppl.: 267–272.

Haldane, JBS (1949b) Disease and evolution, Symposium sui fattori ecologici e genetici della speciazione degli animali, Pallanza 31 luglio-2 agosto 1948, *La ricerca scientifica*, 19, Suppl: 68–76.

Latte, B. (1968). Aspetti medico-sociali delle microcitemie in Sardegna, in *Atti delle giornate di studio sulla microcitemia*, Cagliari, 27–28 Settembre 1968, Istituto Italiano di Medicina Sociale, Roma.

Lehndorff, H. (1936). Die Erythroblastenanämie. *Ergebn. Inn. Me. U. Kinderh*., **50,** 568.

Montalenti, G. (1954). The genetics of microcythemia. *Caryologia*, **4**(Suppl.), 554–588.

Montalenti, G., Silvestroni, E., and Bianco, I. (1953). Nuove indagini sul problema della microcitemia. *Rendiconti dell'Accademia Nazionale dei Lincei*, ser. **8**(14), 183–188.

Montalenti, G., Silvestroni, E. and Bianco, I. (1959). Inchiesta genetico-statistica sulla struttura di popolazioni microcitemiche. I. Aluni dati su un centro di 4000 abitanti con elevata frequenza genica nel ferrarese, *La Ricerca Scientifica* **19**(Suppl.), 119–127.

Neel, J.V. (1951). The population genetics of two inherited blood dyscrasias in man. *Cold Spring Harb. Symp. Quant. Biol*., **15,** 141–158.

Neel, J.V. and Valentine, W.M. (1947) Further studies on the genetics of thalassemia. *Genetics*, **32,** 38–63.

Ortolani, M. (1941). Anemia di Cooley ed altre sindromi eritroblastiche dell'infanzia. *Clinica Pediatrica*, **23,** 45.

Ortolani, M. (1946). La diagnosi di anemia di Cooley. *Gazzetta Medica Italiana*, **104–105**(1), 29–40.

Pachioli, R. (1940). La mielosi eritremica cronica tipo Cooley. *Clinica Pediatrica*, **22,** 233–430.

Paradiso, F. (1942). Razza, ereditarietà e infezione palustre nell'etiologia dell'anemia di Cooley. *Gazzetta Medica Italiana*, **51,** 22–39.

Silvestroni E. e Bianco I. (1943) Prime osservazioni di resistenze globulari aumentate in soggetti sani e rapporto fra questi soggetti e i malati di cosiddetto ittero emolitico con resistenze globulari aumentate. *Bollettino ed Atti della Accademia Medica di Roma*, 69, 11-12: 293–309, seduta del 26 Novembre 1943.

Silvestroni, E. (1949). Microcitemia e malattie a substrato microcitemico. Falcemia e malattie falcemiche. *Atti del 50° Congresso della Società Italiana di Medicina Interna*, Ed. Pozzi, Roma, 1949.

Silvestroni E e Bianco I (1945a) Dimostrazione nell'uomo di una particolare anomalia ematologica costituzionale e rapporti fra questa anomalia e l'anemia microcitica costituzionale. *Policlinico* Sez. Med., 52: 105-137.

Silvestroni, E e Bianco, I. (1945b). Il metodo di Simmel per lo studio delle resistenze globulari. *Policlinico Sez. Prat.*, **51,** 153–158.

Silvestroni E. e Bianco I. (1946) Ricerche sui familiari sani di malati di morbo di Cooley», *Ricerche di Morfologia*, 22: 217–256.

Silvestroni, E. and Bianco, I. (1946–1947). Nuove ricerche sui famigliari di malati di morbo di Cooley e prime osservazioni sulla frequenza dei portatori di microcitemia nel ferrarese e in alcune regioni limitrofe. *Bollettino ed Atti dell'Accademia Medica di Roma*, **72**: 32–33 seduta del 30 Novembre 1946.

Silvestroni, E. and Bianco, I. (1947a). Nuove ricerche sulla trasmissione ereditaria della microcitemia. *Policlinico, Sez. Prat.,* **54,** 1359–1370.

Silvestroni, E. and Bianco, I. (1947b). Sulla frequenza dei portatori della microcitemia nel Ferrarese, sui gruppi sanguigni dei microcitemici e sulla trasmissione ereditaria della microcitemia. *La Ricerca Scientifica*, **17**(12), 2021–2024.

Silvestroni, E. and Bianco, I. (1948a). Sulla frequenza della microcitemia nel Ferrarese e in alcune altre regioni d'Italia. *Policlinico, Sez. Prat.,* **55,** 417–429.

Silvestroni, E. and Bianco, I. (1948b). Nuove ricerche sull'eziologia del morbo di Cooley e prime osservazioni sulla frequenza della microcitemia nel Ferrarese. *Minerva Medica*, **39**(1), 8–21.

Silvestroni, E. and Bianco, I. (1949). Sulla frequenza della microcitemia nel Ferrarese e in alcune altre regioni d'Italia. *Policlinico, Sez. Prat.*, **56,** p. 906.

Silvestroni, E. e Bianco, I., Montalenti, G. and Siniscalco M. (1950). Frequency of Microcytemia in some italian district. *Nature*, **165,** 682.

Silvestroni, E. e Bianco, I., Montalenti, G. and Siniscalco, M. (1954) Genic equilibrium of Microcytemia in some italian district. *Nature*, **173,** 357–359.

Silvestroni, E., Bianco, I.., e and Alfieri, N. (1952). Sulle origini della microcitemia in Italia e nelle altre regioni della terra. *Medicina*, **2,** 187–216.

Siniscalco, M., Bernini, L., Filippi, G., Latte, B., Merra Kahan, P., and Pomelli, S. (1966). Population genetics of haemoglobin variants, thalassemia and glucose-6-phoshate dehydrogenase deficiency, with particolar reference to malaria hypothesis. *Bull. WHO*, **34,** 379–393.

Siniscalco, M., Bernini, L., Latte, B., and Motulsky, A.G. (1961). Favism and thalassemia in Sardinia and their relationship to malaria. *Nature*, **190,** 1179–1180.

Toscano, F. (1941). La sindrome di ittero-anemia di Rietti-Greppi-Micheli. *L'Ospedale Maggiore di Novara*, **9**.

Valentine WN, Neel JV (1944) Hematologic and genetic study of the transmission of thalassemia (Cooley's anemia; mediterranean anemia). *Archives of internal medicine*, 74: 185–196.

Vallisneri, E. (1940). Su trenta casi di sindrome anemica di Cooley. *Policlinico infantile*, **4,** 145–147.

Vecchio, F. (1947). Contributo allo studio della genetica dell'anemia di Cooley. *La Pediatria*, **10–12,** 529–562.

Weatherall, D.J., Clegg, J.B. (2001) *The thalassaemia syndromes*, Oxford, Blackwell Science, 2001 (see the long chapter 1 "Historical perspectives: the many and diverse routes to our current understanding of the thalassaemias, pp. 3–62).

Weatherall, D.J. (2004a). J.B.S. Haldane and the malaria hypothesis. In K.R. Dronamraju (ed), *Infectious Disease and Host–Pathogen Evolution*. Cambridge University Press, Cambridge, UK, pp. 18–36.

Weatherall D.J. (2004b). Thalassaemia: The long road from bedside to genome. *Nat. Rev. Genet.*, **5**, 1–7.

Whipple GH e Bradford WL (1932) Racial and familial anemia of children. *American Journal of Disease of Children*, 33: 336–365.

Whipple GH, Bradford WL. (1936) Mediterranean disease - thalassemia (erythroblastic anemia of Cooley): associated pigment abnormalities simulating hemochromatosis. *Journal of Pediatrics*, 9: 279–311.

Wintrobe M.M. (ed) (1980) *Blood, Pure and Eloquent*, McGraw-Hill, New York.

Wintrobe M.M. (1985) *Haematology - the Blossoming of Science*, Lea & Febiger, Philadelphia.

Wintrobe MM, Mathews E, Pollack R e Dobins BM (1940) A familial hemopoietic disorder in italian adolescent and adults resembling mediterranean disease (Thalassemia). *Journal of American Medical Association*, 114: 1530–1538.

Resistance to Antimalarial Drugs: Parasite and Host Genetic Factors

Rajeev K. Mehlotra and Peter A. Zimmerman

1. Introduction

Drug resistance in malaria is the most formidable obstacle in the fight against the disease since it jeopardizes the most elementary objective of malaria control – reducing suffering and eliminating mortality. Although several mechanisms have been described; an important cause of drug resistance appears to be point mutations in the malaria parasite protein-target genes. Clinically significant resistance to antimalarial agents seems to require the accumulation of multiple mutations. Efforts to circumvent antimalarial drug resistance range from the use of combination therapy with existing agents to studies directed toward discovering novel targets and therapies. However, the contribution of host genetic factors, particularly those associated with antimalarial drug metabolism, has not yet been explored. In addition, the link between *in vivo* drug concentrations and treatment outcome is often inconsistent. Evidence suggests that in a variety of drug-treated illnesses or diseases, metabolism has a major impact on the effectiveness of a drug. Here, to gain some understanding of the role variation in drug metabolism might play in malaria control, we provide background information on global distribution of malaria (Section 2), drugs available to treat malaria (Section 3), and the occurrence of antimalarial drug resistance in association with mutations in specific parasite genes (Section 4). In Section 5, we then turn our attention to variability in structure and function of enzymes involved in Phase I (cytochrome P450) and Phase II uridine diphosphate glucuronosyltransferase (UGT) drug metabolism pathways in the human host. Although relatively few studies have been performed, our review

---

Rajeev K. Mehlotra and Peter A. Zimmerman • Center for Global Health and Diseases, Case Western Reserve University, School of Medicine, Cleveland, Ohio.

Malaria: Genetic and Evolutionary Aspects, edited by Krishna R. Dronamraju and Paolo Arese, Springer, New York, 2006.

will suggest that the variability in antimalarial drug effectiveness is associated with the host genetic polymorphism.

2. Malaria

Malaria, one of the most important parasitic infections, is on the rise globally. This has resulted in an increase in the morbidity and mortality from malaria in endemic areas, a resurgence in areas where it was previously eradicated and an increase in imported malaria in Europe and North America (Maitland *et al.*, 2003; Suh *et al.*, 2004). Over the past 35 years, the incidence of malaria has increased twofold to threefold. At present, malaria occurs in over 100 countries, 300–500 million people experience a malaria infection per year, and up to 2.5 million malaria–attributable deaths occur annually (Suh *et al.*, 2004; Hartl, 2004). The continuing upsurge has come from a coincidence of drug–resistant parasites, insecticide–resistant mosquitoes, global climate change, and socio–economic and political factors (Hartl, 2004). Population projections indicate that approximately 400 million births will occur within malarious regions by 2010. By this date, the Roll Back Malaria initiative is challenged to halve the world's malaria burden (Hay *et al.*, 2004).

Four species of malaria parasite cause disease in humans: *Plasmodium falciparum*, *Plasmodium vivax*, *Plasmodium malariae*, and *Plasmodium ovale*. Of these, *P. falciparum* causes most severe disease and mortality as a result of its prevalence, virulence, and drug resistance. Eighty per cent of falciparum malaria cases occur in tropical Africa, where most people become infected during childhood, and most of the morbidity and mortality are seen in children under the age of 5 years due to severe malaria (Snow *et al.*, 2001). High morbidity and mortality is also seen in young women who are at risk from severe anemia during pregnancy, and may bear low birth-weight babies (Brabin *et al.*, 1990; Steketee *et al.*, 2001). Falciparum malaria infection during pregnancy may also result in stillbirth (Goldenberg and Thompson, 2003) and infant mortality (Guyatt and Snow, 2001). Due to resistance to commonly used drugs, such as chloroquine (CQ) and Fansidar (sulfadoxine–pyrimethamine, SP), *P. falciparum* has also become the predominant species in many parts of the world outside of Africa. Falciparum malaria in South America, the Indian subcontinent, Southeast Asia, and China tend to be less common than in tropical Africa, and all age groups are susceptible to severe malaria (Winstanley, 2000). Falciparum malaria also continues to be the principal medical threat to travelers in tropical zones (Le Mire *et al.*, 2004).

It has been estimated that the global burden of malaria due to *P. vivax* is approximately 70–80 million cases annually, and approximately 10–20% of all the cases occur in sub-Saharan Africa (Mendis *et al.*, 2001). Outside of Africa, *P. vivax* accounts for ≥50% of all malaria cases. About 80–90% of *P. vivax* outside of Africa occurs in the Middle East, Asia, and the western Pacific, mainly in tropical regions, and 10–15% occurs in Central and South America (Mendis *et al.*, 2001). Although the effects of repeated attacks of *P. vivax* through childhood and adult life are seldom fatal, they can have major deleterious effects on personal well-being, growth, and development (Mendis *et al.*, 2001). *Plasmodium vivax* infection can cause anemia (Collins *et al.*, 2003), and, during pregnancy, it is associated with maternal anemia and low birth weight (Nosten *et al.*, 1999).

The other two species, *P. malariae* and *P. ovale*, though less prevalent than *P. falciparum* and *P. vivax*, have a widespread, but patchy, distribution. *Plasmodium malariae* infections are common in the Brazilian Amazon (Scopel *et al.*, 2004), sub-Saharan Africa (Tahar *et al.*, 1998; Tobian *et al.*, 2000), and East/Southeast Asia (Kawamoto *et al.*, 1999; Mehlotra *et al.*, 2000). *Plasmodium ovale* is more commonly found in sub-Saharan Africa (Ukpe, 1998; Lowenthal, 1999; Faye *et al.*, 2002) and East/Southeast Asia (Kawamoto *et al.*, 1999; Win *et al.*, 2002) and is less common in South America (Lysenko and Beljaev, 1969). In addition to these four species, humans naturally infected with the simian parasite *P. knowlesi* have recently been identified in Malaysia (Singh *et al.*, 2004).

3. Antimalarial Chemotherapy and Chemoprophylaxis

Chemotherapy and chemoprophylaxis are the principal means of combating malaria parasite infections in the human host. In the last 70 years, since the introduction of synthetic antimalarials, only a small number of compounds have been found suitable for clinical usage, and this limited armament is now severely compromised by the spread of drug–resistant parasite strains (Winstanley, 2000; Bloland, 2001).

3.1. Drugs Available for Treatment of Malaria

The most widely used drugs have been quinine and its derivatives, and antifolate combination drugs.

3.1.1. Quinine and Related Compounds

Quinine, along with its dextroisomer quinidine, has been the drug of last resort for the treatment of malaria, particularly for severe malaria. Chloroquine, a 4-aminoquinoline derivative of quinine was first synthesized in 1934, and has since been the most widely used antimalarial drug. However, its usefulness has declined in those parts of the world, where strains of *P. falciparum* and *P. vivax* have emerged that are resistant to its action. Amodiaquine, a relatively widely available compound, is a side-chain analog of CQ, and is effective against chloroquine–resistant (CQR) falciparum malaria. Other quinine-related compounds in common use include mefloquine, a 4-quinoline–methanol derivative of quinine, and the 8-aminoquinoline derivative primaquine, specifically used for eliminating the late hepatic stages and latent tissue forms of *P. vivax* and *P. ovale* that cause relapses.

3.1.2. Antifolate Combination Drugs

Antifolate drugs include various combinations of dihydrofolate reductase (DHFR) inhibitors, such as pyrimethamine, proguanil, chlorproguanil, and cycloguanil (an active metabolite of proguanil, structurally related to pyrimethamine), and sulfa drugs, such as sulfadoxine, sulfalene, and dapsone. Though these drugs have antimalarial activity when used alone, resistance can develop rapidly. When used in combination, they produce a synergistic effect on the parasite, and can be effective even in the presence of resistance to the individual components (Bloland, 2001). Typical combinations include sulfadoxine–pyrimethamine (SP or Fansidar) and sulfalene–pyrimethamine (Metakelfin). A new antifolate combination drug Lapdap (chlorproguanil and dapsone) is currently being tested in Africa (Winstanley, 2001; Lang and Greenwood, 2003). The drug is effective against SP–resistant *P. falciparum*, and is rapidly eliminated from the body, giving it low selection pressure for drug resistance (Bloland, 2001). However, SP–resistant parasites can become resistant to Lapdap if a specific mutation is added to the *dhfr* mutant haplotype (see below in Section 4).

3.1.3. Antibiotics

Tetracycline and derivatives, such as doxycycline, are potent antimalarials, and are used for both treatment and prophylaxis. Tetracyclines, in combination with quinine, are particularly useful for the treatment of acute malaria due to multidrug-resistant strains of *P. falciparum* (Bloland, 2001).

3.1.4. Artemisinin and Its Derivatives

A unique class of compounds (ART drugs) – sesquiterpene lactone endoperoxides – originates from the Chinese herb qing hao (*Artemisia annua*), used for centuries to treat malaria and other parasitic diseases

(Price, 2000; Balint, 2001). The parent compound is artemisinin (quinghaosu), from which analogues artesunate, dihydroartemisinin, artemether, and arteether, exhibiting varying pharmacokinetic properties, are derived (Jung *et al.*, 2004; Ploypradith, 2004). These drugs are widely used in China, Southeast Asia, and parts of Africa. The ART compounds, still under various phases of clinical study, have become the first-line drugs for the treatment of multidrug-resistant *P. falciparum* infections in Southeast Asia (Olliaro and Taylor, 2004). The Department of Health, Papua New Guinea (PNG), has recently (2000–2001) approved the use of ART drugs as the second-line drugs of choice for severe/complicated and treatment failure-malaria cases, in combination with Fansidar.

The current high level of interest in ART drugs is due to the following advantageous features. These drugs have large therapeutic windows and, on the basis of extensive human use, appear to be safe, even in children and pregnant women. Furthermore, there is no reported "added toxicity" when ART drugs are used in humans in combination with other compounds (Taylor and White, 2004). These compounds act rapidly upon asexual blood stages of *P. vivax* and CQS, CQR, and multidrug-resistant strains of *P. falciparum*. They reduce the parasite biomass very quickly (by around 4-logs for each asexual cycle), and have the considerable advantage over other antimalarials because they exhibit gametocytocidal action (Kumar and Zheng, 1990; Nosten *et al.*, 1998; White and Olliaro, 1998). Through rapid killing of *Plasmodium* asexual stages and gametocytes, ART drugs limit the transmission of the parasite to new hosts. Furthermore, these activities confer an important theoretical advantage by decreasing the potential for drug resistance mutations to be inherited by offspring parasites (Winstanley *et al.*, 2002).

3.1.5. Miscellaneous Compounds

Halofantrine was developed in the 1960s by the Walter Reed Army Institute of Research. It is a phenanthrene methanol structurally related to quinine. This synthetic antimalarial is effective against multidrug-resistant falciparum malaria (including mefloquine-resistant). The aryl alcohol lumefantrine, similar to mefloquine and halofantrine, is a relatively new antimalarial synthetic molecule that was developed in the 1970s by the Academy of Military Medical Sciences, China. Lumefantrine is also effective against CQR *P. falciparum* isolates. A combination of lumefantrine and artemether (Coartem) has proved to be effective for treating acute, uncomplicated malaria (Omari *et al.*, 2004). Atovaquone is a hydroxynaphthoquinone that is effective against CQR *P. falciparum*. Until recently, a combination of atovaquone and proguanil (Malarone) was considered to be effective for the treatment and prophylaxis of multidrug-resistant falciparum malaria. However, first cases of resistance have just recently been reported from Africa (Wichmann *et al.*, 2004; Giao *et al.*, 2004; Muehlen *et al.*, 2004). Piperaquine is a member of the 4-aminoquinoline group that includes CQ. In the late 1970s, piperaquine replaced CQ as the antimalarial recommended by the National Control Program of China, and has been used since as the first-line treatment for CQR malaria. The piperaquine-based formulation most recently tested in clinical trials in China, Vietnam, and Cambodia is a combination with dihydroartemisinin (Artekin), which holds promise for use in CQR endemic regions (Tran *et al.*, 2004; Karunajeewa *et al.*, 2004; Hung *et al.*, 2004).

4. Antimalarial Drug Resistance

4.1. *Current Status of Drug-resistant Malaria*

In general, four basic approaches have been routinely used to assess antimalarial drug resistance: *in vivo*, *in vitro*, animal models, and molecular characterization. Careful consideration of the type of information each approach yields indicates that these methods are complementary, rather than competing, sources of information about resistance. In the *in vivo* methods, demonstration of persistence of parasites in a patient receiving directly observed therapy is usually considered sufficient to recognize drug resistance. Drug resistance, particularly aminoquinoline resistance, can be graded into different levels depending on the timing of the recrudescence following treatment (Box 5.1).

In vivo resistance to antimalarial drugs has been described for three of the four species of malaria parasite that naturally infect humans, *P. falciparum*, *P. vivax*, and *P. malariae*. *Plasmodium falciparum* has devel-

Box 5.1. Levels of Antimalarial Drug Resistance and Therapeutic Response

RI: Recrudescence of infection between 7 and 28 days of completing treatment following initial resolution of symptoms and parasite clearance.
RII: Reduction of parasitemia by >75% at 48 h, but failure to clear parasites within 7 days.
RIII: Parasitemia does not fall by >75% within 48 h.
Adequate clinical response (ACR): absence of parasitemia (irrespective of fever) or absence of clinical symptoms (irrespective of parasitemia) on day 14 of follow-up.
Early treatment failure (ETF): persistence of clinical symptoms in the presence of parasitemia during the first 3 days of follow-up.
Late treatment failure (LTF): reappearance of symptoms in the presence of parasitemia during days 4–14 of follow-up.

oped resistance to nearly all antimalarials in current use, although the geographic distribution of resistance to any single antimalarial drug varies greatly (Wernsdorfer, 1994; Bloland, 2001; Wongsrichanalai *et al.*, 2002). Quinine resistance occurs in East/Southeast Asia (McGready and Nosten, 1999; Mohapatra *et al.*, 2003), and, although reported, is considered to occur infrequently in South America (Demar and Carme, 2004). Chloroquine-resistant falciparum malaria has been described from almost every region where *P. falciparum* is transmitted except from malarious areas of central America and limited areas of the Middle East and central Asia. Southeast Asia has highly variable distribution of falciparum drug resistance; while some areas have high prevalence of absolute resistance to multiple drugs, elsewhere there is a spectrum of sensitivity to various drugs. Chloroquine-resistant *P. falciparum* is widespread in South America (Cortese *et al.*, 2002; Vieira *et al.*, 2004) and Africa (Bloland, 2001; Winstanley *et al.*, 2002). Due to CQR, an increasing number of countries in Africa shifted to SP treatment as the first-line alternative to CQ. Unfortunately, SP resistance, already reported from Southeast Asia (Wernsdorfer, 1994; Wongsrichanalai *et al.*, 2002; Nair *et al.*, 2003) and South America (Wernsdorfer, 1994; Cortese *et al.*, 2002), is becoming more prevalent in Africa (Winstanley *et al.*, 2002; Wongsrichanalai *et al.*, 2002). Reports of amodiaquine resistance have started appearing from South America (Gonzalez *et al.*, 2003), Southeast Asia (Lopes *et al.*, 2002), and Africa (Basco *et al.*, 2002; Ochong *et al.*, 2003; Pettinelli *et al.*, 2004). Mefloquine resistance is frequent in some areas of Southeast Asia (Mockenhaupt, 1995; Pickard *et al.*, 2003), Africa (Mockenhaupt, 1995; Wichmann *et al.*, 2003), and has been reported in the Amazon region of South America (Cerutti *et al.*, 1999). Cross-resistance (tolerance to a usually toxic substance as a result of exposure to a similarly acting substance) between mefloquine and halofantrine is suggested by reduced response to halofantrine, when used to treat mefloquine-resistant cases (Ketrangsee *et al.*, 1992; ter Kuile *et al.*, 1993; Wilson *et al.*, 1993).

It has been found that if a drug, for which the parasite has developed resistance, is withdrawn from use for some time, or if it is combined with another effective drug, the sensitivity to that drug may return. In 1993, Malawi stopped the use of CQ to treat falciparum malaria cases because of high treatment failure rate (~80%), and replaced CQ with SP. By 2000–2001, this switch in recommended antimalarial treatment policy appeared to have resulted in the return of CQ efficacy to nearly 100% in two areas of Malawi (Laufer and Plowe *et al.*, 2004). An increased prevalence of the CQS wild type *pfcrt* haplotype (see the following subsection) in *P. falciparum* populations resulted in the recovery of CQ sensitivity (Mita *et al.*, 2004). In certain areas of Southeast Asia where mefloquine resistance prevails, the combination of an ART drug and mefloquine has improved mefloquine sensitivity (Brockman *et al.*, 2000). On the other hand, in a recent *in vivo* efficacy study of CQ plus artesunate for the treatment of symptomatic malaria in children in Guinea–Bissau, neither a beneficial nor a detrimental effect of the two drugs was observed (Kofoed *et al.*, 2003). In another study involving children with uncomplicated malaria in Burkina Faso, it was found that although artesunate improved the parasite clearance, 51% of the cases still failed to clear parasites on day 28 post-treatment (Sirima *et al.*, 2003).

Plasmodium vivax infections acquired in some areas have been shown to be resistant to CQ and/or primaquine (Kain, 1995; Collins and Jeffery, 1996). Chloroquine-resistant vivax malaria has been reported from PNG and Indonesia (Baird *et al.*, 1991, Murphy *et al.*, 1993; Baird *et al.*, 1996) and from India (Dua *et al.*, 1996). Emergence of CQR vivax malaria has been reported from Myanmar and Vietnam (Marlar-Than *et al.*, 1995; Phan *et al.*, 2002). In certain areas of Thailand, CQ is still effective against *P. vivax* infections (Looareesuwan *et al.*, 1999); however, primaquine resistance (Wilairatana *et al.*, 1999), mefloquine resistance (Chotivanich *et al.*, 2004), and SP resistance (Wilairatana *et al.*, 1999) have been observed. Recently, *P. vivax* infections resistant to SP therapy have been observed in Indonesia (Hastings *et al.*, 2004). Chloroquine and/or primaquine-resistant *P. vivax* infections are prevalent in South America (Garavelli and Corti, 1992; Phillips *et al.*, 1996; Soto *et al.*, 2001). Primaquine-resistant *P. vivax* infections have also been reported from sub-Saharan Africa (Smoak *et al.*, 1997; Schwartz *et al.*, 2000).

Recently, CQR *P. malariae* has been reported from southern Sumatra, Indonesia (Maguire *et al.*, 2002); however, the presence of CQR *P. malariae* in nearby PNG was indicated earlier by Desowitz and Spark (1987).

Artemisinin and its derivatives represent a valuable class of antimalarials because these rapidly acting drugs are effective against severe malaria and multidrug-resistant *P. falciparum*. However, their increasing and unregulated use is a risk factor for the emergence of drug resistance. Clinically relevant ART drug resistance is not yet apparent, however, reduced *in vitro* sensitivity of *P. falciparum* isolates to these drugs has been seen, and malaria cases showing early/late treatment failure with delayed parasite clearance time have been reported. Clinical isolates and laboratory strains of *P. falciparum* vary in their sensitivities to ART drugs. In several *in vitro* studies, IC_{50} values (inhibitory concentration of a drug required to kill 50% of the parasites) for ART drugs were 1.0–10.0 nM (Brossi *et al.*, 1988; Shmuklarsky *et al.*, 1993 Bustos *et al.*,

1994; Gay *et al.*, 1994; Wongsrichanalai *et al.*, 1997; Ringwald *et al.*, 1999). Artemisinin-resistant mutants of *P. falciparum*, showing a 3- to 10-fold increase in IC_{75} and IC_{95} values of the drug, were generated in the laboratory (Inselberg, 1985). Innate, transient resistance to artemisinin was observed in four out of 50 *P. falciparum* isolates from Nigeria (Oduola *et al.*, 1992). IC_{50} values of the drug for these isolates were 7- to 14-fold higher than the reference strain when evaluated 1 week after culture adaptation. All four isolates became susceptible to artemisinin after 3 weeks of cultivation. A *P. falciparum* strain, isolated from a patient from Mali, showed IC_{50} values of ART drugs above the 90th percentile (IC_{50} value was 2.3-fold higher for arteether) (Gay *et al.*, 1994). During the period 1988–1999, *P. falciparum* parasites in Yunnan Province, China, showed a 3.3-fold increase in their artesunate IC_{50} values, indicating decreased *in vitro* drug sensitivity (Yang *et al.*, 2003). In an *in vitro* sensitivity study of *P. falciparum* isolates from Madagascar, four out of 51 isolates exhibited IC_{50} values between 12 and 17.5 nM, indicative of resistance to this drug (Randrianarivelojosia *et al.*, 2001). In Thailand (Luxemburger *et al.*, 1998; Treeprasertsuk *et al.*, 2000) and Vietnam (Le *et al.*, 1997; Giao *et al.*, 2001), cases of falciparum malaria treated with ART drugs have shown early treatment failure with delayed parasite clearance time (up to 17 days). Two patients exhibited RII response (Giao *et al.*, 2001).

Monotherapy with ART drugs leads to high recrudescence, possibly due to suboptimal drug levels, rapid drug clearance, and patient noncompliance (Hyde, 2002). This recrudescence may not be the result of inherent parasite resistance, and could be due to other factors, such as initially high parasite burdens (Ittarat *et al.*, 2003). It has also been suggested that some parasites respond to the drug pressure by entering a dormant state that is resistant to further inhibition (Hoshen *et al.*, 2000). It is, therefore, preferential to administer a combination of an ART drug with a longer acting drug of a different family, to increase killing of any residual parasites (Peters, 1998). Combining an ART drug with another efficacious antimalarial drug is increasingly viewed as the optimal therapeutic strategy for malaria (Olliaro and Taylor, 2004; Nosten and Ashley, 2004; White, 2004). However, further investigations are needed to ascertain its success in diverse malaria–endemic regions (Bloland *et al.*, 2000).

4.2. Parasite Genetic Polymorphism As a Basis for Antimalarial Resistance

Drug resistance in *P. falciparum* arises through the selection of mutations associated with reduced drug sensitivity. Drug-resistant parasites are more likely to be selected if parasite populations are exposed to subtherapeutic drug concentrations through (1) unregulated drug use, (2) use of inadequate drug regimens, and (3) the use of long half-life drugs (Wernsdorfer and Payne, 1991; Talisuna *et al.*, 2004). Recently, significant progress has been made to understand the molecular mechanisms underlying drug resistance in *P. falciparum* (Hyde , 2002; Le Bras and Durand, 2003; Wernsdorfer and Noedl, 2003). Linkage analysis has been useful in locating drug resistance genes in *P. falciparum* (Hayton and Su, 2004; Sen and Ferdig, 2004; Anderson, 2004).

4.3. Genes Associated with Chloroquine Resistance in P. *falciparum*

It is thought that CQR in *P. falciparum* arose independently in Southeast Asia and South America during the late 1950s and spread to the rest of the world (Payne, 1987). Since CQR took many years to evolve, it is proposed to involve greater genetic complexity than other examples of antimalarial drug resistance (e.g., pyrimethamine resistance, described later). Instead of a single gene–single mutation scenario, it has been suggested by several investigators that multiple mutations in multiple genes underlie the basis of CQR (Chen *et al.*, 2002; Mu *et al.*, 2003).

The phenotypic relationship of drug efflux and resistance reversal by verapamil, a calcium-channel blocker, between mammalian *m*ulti*d*rug *r*esistance and CQR was exploited to identify a homologue gene (*pfmdr1*, chromosome 5) in CQR *P. falciparum* isolates (Foote *et al.*, 1990; Volkman and Wirth, 1998). A P-glycoprotein homologue 1 (Pgh1), encoded by the *pfmdr1*, was found to be located at the parasite digestive vacuole membrane (Cowman *et al.*, 1991). In the CQS isolates, the *pfmdr1* "allele" was predicted to carry amino acids N-Y-S-N-D (Table 5.1). Among CQR isolates, 1-4 key nucleotide differences resulted in amino acid substitutions. There were two CQR-associated alleles (Table 5.1).

The mutation 184F was considered not to be involved in CQR as it occurred in CQS isolates as well (Foote *et al.*, 1990). In a subsequent study, *pfmdr1* gene amplification was found to be associated with a decrease in CQR, increased mefloquine resistance, and cross-resistance to halofantrine and quinine. However, amodiaquine sensitivity remained unchanged (Cowman *et al.*, 1994). Later, in order to examine the role of S1034C, N1042D, and D1246Y mutations of Pgh1 in controlling the parasite sensitivity to various antimalarials, parasite transfection and allelic exchange studies were performed (Reed *et al.*, 2000). From these studies, it was concluded that polymorphisms in *pfmdr1* alone were insufficient to confer CQR, but confer quinine resistance and markedly

Table 5.1. Geographic Distribution of *pfmdr-1* Alleles

		Codon				
Region/country	CQ phenotype	86	184	1034	1042	1246
Worldwide	Sensitive	N	Y	S	N	D
Southeast Asia+ and Africa	Resistant	Y	Y	S	N	D
South America	Resistant	N	F	C	D	Y
Colombia*	?	N	F	S	D	Y
Papua New Guinea*	?	N	F	S	N	D
	?	Y	F	S	N	D

+: Including Papua New Guinea.
*: Mehlotra and Zimmerman (unpublished).
?: Not yet characterized.

affect the parasite susceptibility to mefloquine, halofantrine, and artemisinin (Reed *et al.*, 2000). In another study, involving a genetic cross, it was found that *pfmdr1* sequence polymorphisms are important determinants of sensitivity to mefloquine, halofantrine, lumefantrine, artemisinin, artemether, and arteflene (another derivative of artemisinin) (Duraisingh *et al.*, 2000).

In parallel studies searching for the CQR genetic locus in *P. falciparum*, using the progeny from cross between CQS HB3 and CQR Dd2 strains, CQR phenotype was found to be associated with a ~400 kb region on chromosome 7 (Wellems *et al.*, 1991). This ultimately led to identification of the *pfcrt* (*P. falciparum* chloroquine *r*esistance *t*ransporter) gene (Fidock *et al.*, 2000). The coding region of *pfcrt* spans 3.1 kb in 13 exons. The PfCRT protein was found to be located at the parasite digestive vacuole membrane (Fidock *et al.*, 2000), and may be involved in drug flux and/or pH regulation in an allele-specific manner (Carlton *et al.*, 2001; Bennett *et al.*, 2004). Eight point mutations (M74I, N75E, K76T, A220S, Q271E, N326S, I356T, and R371I) in *pfcrt* distinguished CQR from CQS progeny of the HB3 X Dd2 cross. Based on a survey of parasite strains collected from around the world, it was observed that a number of distinct *pfcrt* haplotypes are associated with CQR in different malarious regions (Fidock *et al.*, 2000). Initially, the following *pfcrt* "alleles" (amino acids 72–76) were characterized: CQS–CVMNK (all regions), CQR–CVIET (Asia and Africa), SVMNT, CVMNT, and CVMET (South America) (Fidock *et al.*, 2000). Since then, new *pfcrt* alleles have been found (Table 5.2).

As the K76T mutation is observed in all CQR parasite lines, it was concluded that this mutation is critical to acquire the CQR phenotype. Subsequently, clones were generated *in vitro* from a CQS parasite line

Table 5.2. Geographic Distribution of *pfect* Alleles

Region/country	CQ phenotype	Codon				
		72	73	74	75	76
Worldwide	Sensitive	C	V	M	N	K
Southeast Asia, Africa, and South America	Resistant	C	V	I	E	T
South America, Southeast Asia, and India	Resistant	S[1]	V	M	N	T
South America and Southeast Asia	Resistant	C	V	M	N	T
Colombia	Resistant	C	V	M	E	T
South America	Resistant	S[2]	V	M	N	T
Indonesia	Resistant	C	V	I	K	T
	Resistant	S	V	I	E	T
Cambodia	Resistant	C	V	I	D	T
	Resistant	C	V	T	N	T
Guyana	Resistant	S	V	M	I	T
	Resistant	R	V	M	N	T

1: S[agt]VMNT; 2: S[tct]VMNT.

containing CVIEK allele. These clones showed novel K76N or K76I mutation, and exhibited 8- and 12-fold higher CQ IC_{50} values, respectively, than that of the original parasite line (Cooper *et al.*, 2002). These results suggested that the loss of lysine is central to the CQR mechanism. To address whether mutations in *pfcrt* are sufficient to confer CQR, Sidhu *et al.* (2002) performed allelic exchange manipulations to replace the endogenous *pfcrt* allele of a CQS line (CVMNK allele) with *pfcrt* alleles from CQR lines (CVIET, SVMNT, or CVIEI allele). Although phenotypic analyses showed that mutant *pfcrt* alleles conferred CQR phenotype to the transfected CQS parasite line, the possibility that other loci can contribute to this phenotype was not ruled out. Phenotypic characterization also revealed increased susceptibility (twofold to fourfold) to quinine, mefloquine, and artemisinin in the *pfcrt*-transformed clones. However, the amodiaquine sensitivity remained unchanged in the *pfcrt*-transformed clones (Sidhu *et al.*, 2002). Recently, using the allelic exchange approach, it has been shown that CQR *pfcrt* haplotypes can generate an increase in susceptibility (threefold) to halofantrine and mefloquine, and can also confer greater sensitivity (10-fold) to the antiviral agent amantadine (Johnson *et al.*, 2004).

Recently, quinine resistance has been found to be linked to polymorphisms in *pfnhe-1* gene, located on chromosome 13, which encodes a putative Na^+/H^+ exchanger (Ferdig *et al.*, 2004).

4.4. Mechanism of Resistance to Antifolate Combination Drugs in *P. falciparum*

4.4.1. DHFR Inhibitors

It has been shown that the major mechanism of resistance to pyrimethamine (PYR) in *P. falciparum* is due to altered drug binding to DHFR (Chen *et al.*, 1987; Zolg *et al.*, 1989). This suggested that the protein was structurally altered, and was consistent with the idea that mutations in the DHFR enzyme decreased the affinity of PYR binding. Cloning of the *dhfr–ts* gene (*P. falciparum* chromosome 4) has allowed detailed structural and functional analyses of the enzyme (Bzik *et al.*, 1987; Cowman *et al.*, 1988; Peterson *et al.*, 1988; Snewin *et al.*, 1989), and revealed that high level PYR resistance results from accumulating mutations in the *dhfr* domain at codons S108N, N51I and C59R (Sibley *et al.*, 2001; Yuthavong, 2002). More recently, transfection of the *dhfr* gene containing 108N mutation or 108N/51I/59R mutations into drug-sensitive *P. falciparum* conferred the predicted level of PYR resistance (Wu *et al.*, 1996). Amino acid positions 16 and 164 (A16V and I164L), as well as an alternative mutation at position 108 (S108T), are also involved in resistance to DHFR inhibitors, particularly in the case of cycloguanil, the active metabolite of proguanil (Foote *et al.*, 1990; Peterson *et al.*, 1990; Hyde, 1990). Finally, it has been found that addition of the 164L mutation to the 108N/51I/59R *dhfr* allele confers not only high level resistance to PYR, but also to chlorproguanil, the DHFR inhibitor used in the combination drug Lapdap (Wichmann *et al.*, 2003).

4.4.2. Sulfa Drugs

It was found that in sulfadoxine (SDX)-resistant *P. falciparum*, metabolism of the drug was reduced (Dieckmann and Jung, 1986), suggesting that mutations in the dihydropteroate synthase (DHPS) enzyme might be involved in the mechanism of resistance to this class of antimalarials. Sequence analysis of the *dhps* gene (*P. falciparum* chromosome 8) from SDX-resistant and SDX-sensitive strains of *P. falciparum* identified four amino acid differences at positions 436, 437, 581, and 613 (Triglia and Cowman, 1994). Later studies revealed that variability at position 540 was also commonly observed in field samples (Plowe *et al.*, 1997; Triglia *et al.*, 1997; Wang *et al.*, 1997). Mutations at position 436 are: S436A, S436F, and S436C; at position 613: A613S and A613T; and at other positions are: A437G, K540E, and A581AG (Hyde, 2002). A tight linkage between mutations at 436, 437, 581, and 613 and SDX resistance phenotype was observed in the progeny of a genetic cross between SDX-sensitive HB3 and SDX-resistant Dd2 parents (Wang *et al*., 1997). Transfection of the *dhps* gene containing mutations, either singly or in combination, at 436, 437, 540, 581, and 613 into SDX-sensitive *P. falciparum* conferred different levels (5- to 24-fold) of SDX resistance (Triglia *et al*., 1998).

The relationship between mutations in the DHFR and DHPS and clinical SP resistance has recently been assessed on a large scale across diverse geographic areas (Nagesha *et al*., 2001; Contreras *et al*., 2002; Bwijo *et al*., 2003; Anderson *et al*., 2003; Ahmed *et al*., 2004). These studies have found that the DHFR 108N and DHPS 540E mutations were correlated with increased SP resistance. The DHFR 51I, 59R and DHPS 437G mutations were also correlated with resistance to this combination. These data support the idea that mutations in both DHFR and DHPS are important in the mechanism of resistance to the SP combination in the field, and that increasing resistance to this combination is due to progressive accumulation of mutations in these enzymes. In search of a putative explanation for the observed synergy between PYR and SDX, it was found that the status of the *dhps* gene becomes important once mutations have occurred in *dhfr* to a degree where inhibition of DHFR alone by normal therapeutic levels of PYR is insufficient in itself to kill the parasites (Sims *et al*., 1999). In practice, it appears that the triple mutant *dhfr* allele (108N/51I/59R), combined with at least a double mutant *dhps* allele (437G with either 540E or 581G), is associated frequently with SP failure. Recently, in a transfection study using a full-length *dhps* domain mutated at two active-site residues resulting in <10% of the normal activity, it was shown that in transfectants the synergy of the combination PYR/SDX was abolished, proving the role of DHPS in antifolate drug synergy (Wang *et al*., 2004).

4.5. Mechanism of Atovaquone Resistance in *P. falciparum*

Atovaquone is a structural analog of coenzyme Q (ubiquinone) in the mitochondrial electron transport chain. It is the major active component of a new antimalarial combination drug Malarone (atovaquone plus proguanil). The mechanism by which *P. falciparum* develops resistance to this drug is thought to involve point mutations in the parasite cytochrome *b* gene (Korsinczky *et al*., 2000). One of these mutations (Y268S) was detected in the parasites isolated from patients with recrudescence following atovaquone-PYR treatment (Korsinczky *et al*., 2000) and Malarone treatment (Schwartz *et al*., 2003). In another case with recrudescence following Malarone treatment, mutation Y268N was observed (Fivelman *et al*., 2002). However, recently, Malarone treatment failure cases not associated with cytochrome *b* codon 268 mutations have been reported (Wichmann *et al.*, 2004; Farnert *et al.*, 2003), suggesting that other mechanisms may also be involved.

4.6. Mechanisms of Drug Resistance in P. vivax

Mutations in the *P. vivax* orthologue of the *pfcrt* are not associated with CQR (Nomura *et al.*, 2001). This finding suggests a genetic basis for CQR in *P. vivax*, which is different from that in *P. falciparum*. Recently, the *P. vivax* orthologue of the *pfmdr1* (*pvmdr1*) has been identified, and mutations Y976F and F1076L were found in *P. vivax* isolates from different areas of endemicity (Brega *et al.*, 2005). However, the role of these mutations in clinical CQR has not been evaluated in the infected patients. Mutations in the *pv–dhfr* gene, S58R and S117N (equivalent to mutations C59R and S108N in *P. falciparum*) (de Pecoulas *et al.*, 1998a, 1998b; Leartsakulpanich *et al.*, 2002), F57L, S58R and S117N (Imwong *et al.*, 2001), and S117T and other mutations (Imwong *et al.*, 2003; Hastings *et al.*, 2004) have been implicated in clinical PYR resistance. Recently, SDX resistance in *P. vivax* was found to be associated with a specific amino acid, V585 (equivalent to A613 in *P. falciparum*), in the *pv–dhps* gene (Korsinczky *et al.*, 2004).

It is clear that mutations in the parasite genes *pfcrt* and *pfmdr1* play an important role in CQR, and mutations in the *dhfr* and *dhps* genes confer resistance to antifolate combination drugs. However, the association between mutations in these genes and drug response phenotype is not absolute. This is suggested by (a) a wide range of *in vitro* IC_{50} values exhibited by parasites carrying the same mutation. For example, CQ IC_{50} values of 146 *P. falciparum* isolates from various malaria endemic countries were as follows: 8–40 nM for isolates bearing the CQS K76 mutation, 40–60 nM for isolates bearing either K76 or the CQR T76 mutation, and 60–572 nM for isolates bearing T76 mutation (Durand *et al.*, 2001), (b) the observation that CQR *P. falciparum* infections can be cleared by CQ, suggesting that host immunity plays a critical role in the clearance of resistant parasites (Wellems and Plowe, 2001; Djimde *et al.*, 2003), and (c) substantial variations in the metabolism and pharmacokinetics of antimalarial drugs in the human host (see Section 5). Variations in the host drug metabolism may cause subtherapeutic levels of an antimalarial and its active metabolite in the blood, which can cause treatment failure. With this in mind, it is possible that an otherwise sensitive parasite infection may not be cleared due to therapeutically inadequate levels of drug/metabolite. Furthermore, the subtherapeutic drug/metabolite levels are thought to increase the risk of resistance by creating a predisposing environment.

5. Human Drug Metabolism

The liver is the primary site for drug and xenobiotic metabolism, as it contains high concentrations of enzymes necessary for two metabolic pathways: Phase I functionalization reactions and Phase II biosynthetic or conjugation reactions (Goodman and Gilman, 2001). Typical examples of Phase I metabolism include oxidation and hydrolysis. The enzymes involved in Phase I reactions are primarily located in the endoplasmic reticulum of the liver cell and are called microsomal enzymes. The most common Phase I reactions are oxidative processes that involve cytochrome P450 (CYP) enzymes. These enzymes comprise a superfamily of proteins found in almost all living organisms. These enzymes catalyze many reactions: N- and O-dealkylations, aromatic and aliphatic hydroxylations, N- and S-oxidations, and deamination. They are also involved in a number of reductive reactions, generally under oxygen-deficient conditions (Tredger and Stoll, 2002; Guengerich, 2003).

Phase II conjugation introduces hydrophilic functionalities, such as glucuronic acid, sulfate, glycine, or acetyl group, onto drug or drug metabolite molecules. These reactions are catalyzed by a group of enzymes called transferases. Most transferases are located in the cytosol, except UGT that facilitates glucuronidation, which is a microsomal enzyme. Glucuronic acid contains a number of hydroxyl groups and one carboxyl group. This molecule improves the hydrophilicity of a drug/metabolite molecule for rapid renal excretion (Guillemette, 2003; Burchell, 2003).

5.1. Cytochrome P450 Enzyme Superfamily

Cytochrome P450 genes have been found in animals, plants, yeast, and bacteria, with more than 1000 individual *CYP* genes identified in all species studied. For a current list, please consult Nelson D, 2003, http://drnelson.utmem.edu/CytpchromeP450.html. The human *CYP* genes are classified into 17 families. These families contain 57 putatively functional genes, and, in addition, 58 pseudogenes (Guengerich, 2003; Nelson *et al.*, 2004). Of these, four families (designated CYPs 1–4) are involved in the metabolism of drugs; enzymes in the families 1–3 mediate the metabolism of 70–80% of all clinically used drugs (Tredger and Stoll, 2002; Ingelman-Sundberg, 2004a). Seven isoforms in these families (see below) (CYP1A2, CYP2B6, CYP2C9, CYP2C19, CYP2D6, CYP2E1, and CYP3A4) mediate metabolism of most drugs in common use, and three of these – CYP2C9, CYP2C19, and CYP2D6 – mediate ~40% of the total CYP-mediated drug metabolism (Caraco, 1998; Tredger and Stoll, 2002; Ingelman-Sundberg, 2004a).

Each *CYP* gene family is divided into subfamilies (designated by a letter) and, further, into isoforms (designated by a final Arabic number). For example, *CYP2B6* describes a *CYP* family-2 member in the B subfamily, and is isoform 6. Each isoform is the product of a separate gene (Nagata and Yamazoe, 2002; Nelson *et al.*, 2004). Members of the different *CYP* gene families 1–4 share up to 40% sequence homology. Members of the same subfamilies share >55% sequence homology. The genes encoding isoforms within the same subfamily share considerably higher sequence homology; sometimes the DNA sequences are >95% identical (McKinnon , 2000; Tredger and Stoll, 2002).

All genes encoding CYP isoforms in the families 1–3 are polymorphic (Ball and Borman, 1997; Evans and Relling, 1999; Weinshilboum, 2003). The functional importance of the variant alleles differs, and the frequencies of their distribution in different ethnic groups vary (Ingelman-Sundberg, 2004a, 2004b). Table 5.3 provides a general overview of the major CYP isoforms, chromosomal locations of genes encoding them, and the number of single nucleotide polymorphisms associated with amino acid changes; updated information can be found on the Human CYP allele Nomenclature Website http://www.imm.ki.se/CYPalleles/.

These polymorphisms are known to affect the response of individuals to drugs used in treating a variety of illnesses, such as depression, psychosis, cancer, cardiovascular disorders, ulcers and gastrointestinal disorders, pain, and epilepsy (Destenaves and Thomas, 2000; Ingelman-Sundberg, 2004a). Based upon the rate of metabolism for a given drug, four phenotypes can be identified: poor metabolizers (PMs), intermediate metabolizers (IMs), extensive metabolizers (EMs), and ultraextensive metabolizers (UEMs) (Goodman and Gilman, 2001; Ingelman-Sundberg, 2004a; Linder *et al.*, 1997). Genotypically, PMs have both alleles mutated and/or deleted, resulting in lack of the enzyme activity. IMs are heterozygotes for one deficient allele or carry two alleles that cause reduced

Table 5.3. Major Human Drug Metabolizing CYP Isoforms: Their Gene Locations, and Polymorphisms

Isoform	Chromosome	SNPs
CYP1A2	15	11
CYP2B6	19	11
CYP2C9	10	10
CYP2C19	10	11
CYP2D6	22	25–30
CYP2E1	10	3
CYP3A4	7	15

SNPs: Single nucleotide polymorphisms associated with amino acid changes.

enzyme activity. EMs carry two normal alleles. UEMs have multiple gene copies, resulting in increased drug metabolism. The possible consequences of the two extreme phenotypes PM and UEM are explained in Box5.2. (Weinshilboum, 2003; Miller *et al.*, 1997; Ingelman-Sundberg, 2001). The reported frequencies of these extreme phenotypes in various populations for a variety of drugs are: PM 1–23% and UEM 1–30% (Goodman and Gilman, 2001; Nagata and Yamazoe, 2002).

5.2. Uridine Diphosphate Glucuronosyltransferase Enzyme Superfamily

Uridine diphosphate glucuronosyltransferase (UGT) enzymes comprise a superfamily of key proteins that catalyze the transfer of the glucuronic acid group of uridine diphosphoglucuronic acid to the functional group (e.g., hydroxyl, carboxyl, amino, and sulfur) of a specific substrate (Guillemette, 2003; Burchell, 2003). Glucuronidation is one of the major Phase II reactions and accounts for ~35% of all drugs metabolized by Phase II reactions (Evans and Relling, 1999). This biochemical process increases the polarity of the target compounds and facilitates their excretion in urine or bile (Miners and Mackenzie, 1991). It is also involved in protection against environmental toxicants, carcinogens, dietary toxins and regulates homeostasis of numerous endogenous molecules, including bilirubin, steroid hormones, and biliary acids (Wells *et al.*, 2004). Unlike most Phase II reactions that are localized in the cytosol, UGTs are microsomal enzymes (Meech and Mackenzie, 1997).

It has been estimated that in animals, plants, yeast, and bacteria there are at least 110 distinct *UGT* genes whose protein products all contain a characteristic 29 residue carboxy terminus "signature sequence" and, thus, are regarded as members of the same superfamily. Comparison of a relatedness tree of proteins defines 33 families (Guillemette, 2003). The guide-

Box 5.2. Polymorphism in Drug Metabolism: When Does It Really Matter?

Failed efficay
- Metabolite is responsible for activity and individual is PM
- Parent drug must reach threshold and individual is UEM

Increased toxicity
- Parent drug is toxic and individual is PM
- Metabolite is toxic and individual is UEM

For each drug
- Must identify the specific enzyme
- Whether parent drug/metabolite is active/toxic

lines for naming human *UGT* genes are similar to the *CYP* nomenclature system. It is recommended that the root symbol *UGT* be followed by an Arabic number representing the family, a letter designating the subfamily, and an Arabic numeral denoting the isoform within the subfamily, e.g., human *UGT2B7* (Mackenzie *et al.*, 1997; Bock, 2003) (see also UGT homepage http://som.flinders.edu.au/FUSA/ClinPharm/UGT).

In humans, the UGTs are encoded by a multigene family that comprises more than 26 genes. Eighteen of these UGTs correspond to functional proteins and are encoded by two gene families, *UGT1* and *UGT2*, with 41% amino acid similarity between the two families (Guillemette, 2003). UGT1 has one subfamily UGT1A. UGT2 has two subfamilies, UGT2A and UGT2B, with 59% amino acid similarity between them (Guillemette, 2003). The entire UGT1 family is derived from a single, complex gene locus spanning ~210 kb on chromosome 2 (2q37) that is composed of 17 exons. The locus includes 13 unique exon-1 sequences followed by four common exons (exon-2 to exon-5). Of the 13 exon-1 sequences, nine may be spliced and joined with the four common exons, resulting in nine proteins with different N-terminal portions but identical C-terminal portions. These are UGT1A1 and 1A3-1A10. The remaining four exon-1 sequences are considered pseudogenes (p) (UGT1A2p, UGT1A11p, UGT1A12p, and UGT1A13p). Each exon-1 sequence encodes the substrate binding site and is regulated by its own promoter (Ritter *et al.*, 1992; Owens and Ritter, 1995; Gong *et al.*, 2001). In contrast to the UGT1A subfamily, the UGT2B subfamily comprises several independent genes (each composed of six exons) and numerous homologous pseudogenes, all of which are clustered on chromosome 4 (4q13). Similar to the UGT1A subfamily, members of the UGT2B subfamily share the highest degree of divergence in sequences encoded by exons 1 and a high degree of similarity in the C-terminal portion of the protein (Monaghan *et al.*,1994; Beaulieu *et al.*, 1997; Turgeon *et al.*, 2000).

To date, a number of polymorphisms have been described for both *UGT1A* (*UGT1A1* and *1A5-1A10*) and *UGT2B* (*UGT2B4*, *UGT2B7*, *UGT2B15*, and *UGT2B28*) genes (Mackenzie *et al.*, 2000; Miners *et al.*, 2002). Genetic variation may cause different phenotypes by affecting expression levels (Mackenzie *et al.*, 2003) or activities (Little *et al.*, 1999) of UGT enzymes. Variation in UGT expression and/or activity can result in a functional deficit affecting endogenous metabolism, and may lead to jaundice and other diseases (Burchell, 2003). Deficient glucuronidation of drugs and xenobiotics have an important pharmacological impact, which may lead to drug-induced adverse reactions and even cancer (Guillemette, 2003).

5.3. Antimalarial Drug Metabolism

Although antimalarial compounds have been in use since 1930s, until recently, no data were available on the metabolic pathways of these drugs as bioanalytical methods were not available to evaluate absorption–distribution–metabolism–excretion parameters. In recent years, pharmacokinetic properties of some antimalarials have been studied in humans (Krishna and White, 1996). Because chemotherapy is still the principal means to control this disease, this kind of information is needed to optimize antimalarial use. When considered along with the use of drug combinations, it may be possible to increase therapeutic efficiency and evade the emergence of drug-resistant parasite strains (White, 1999).

There have been many methodological developments in the study of drug metabolism over the past 10 years. The increased availability of human tissue (liver microsomes, slices, and hepatocytes) and recombinantly expressed human enzymes are providing insight into how humans metabolize drugs (as reviewed in Masimirembwa *et al.*, 2001). The identification of CYP-specific marker reactions, antibodies, and chemical inhibitors selective for some CYPs have improved the capacity to evaluate the role of different enzymes in the metabolism of test compounds (Rodrigues, 1999; Stormer *et al.*, 2000; Bapiro *et al.*, 2001, 2002a). Through these advancements, much of the pharmacokinetics and metabolic pathways of antimalarial drugs have been elucidated, including the role of CYP enzyme complex (Table 5.4). However, little emphasis has been placed on understanding the basic mechanisms responsible for the pharmacokinetic and pharmacodynamic behaviors of major classes of antimalarial drugs in various populations.

The major metabolic pathway of quinine has been shown to be 3-hydroxylation mediated mainly by CYP3A4 (Zhao *et al.*, 1996). Since quinine caused significant increase in CYP1A enzyme activity and CYP1A1 mRNA expression in HepG2 cells (Bapiro *et al.*, 2002b) and, using recombinant enzymes, CYP1A1 was found to metabolize quinine

Table 5.4. Human CYPs Involved in Antimalarial Drug Metabolism

Antimalarial drug	Primary isoform	Major metabolite	Chromosome
Quinine, Quinidine	CYP3A4	3-Hydroxy derivative	7
Chloroquine	CYP2C8, CYP3A4/3A5	Desethyl derivative	10, 7
Amodiaquine	CYP2C8	Desethyl derivative	10
Mefloquine, Primaquine	CYP3A4, CYP1A2	Carboxy derivative	7, 15
Halofantrine, Lumefantrine	CYP3A4/3A5	Desbutyl derivative	7
Dapsone	CYP2C9, CYP3A4	Hydroxylamine derivative	10, 7
Proguanil	CYP2C19	Cycloguanil	10

rapidly (Li *et al.*, unpublished observations, cited in Bapiro *et al.*, 2002b), it has been suggested that CYP1A1 may also play some role in the metabolism of quinine (Wanwimolruk *et al.*, 1995; Zhang *et al.*, 1997). Similar to quinine, the most important of quinidine's metabolites is 3-hydroxy-quinidine, although quinidine-N-oxide is also formed. CYP3A4, expressed heterologously in yeast (Wrighton *et al.*, 1990) or in human liver microsome preparations (Nielsen *et al.*, 1999), has been shown to metabolize quinidine actively. Although quinidine is a potent inhibitor of CYP2D6 activity (Abraham and Adithan, 2001), the role of this enzyme in the metabolism of quinidine is not clear.

In humans, CQ is metabolized through the N-dealkylation reaction by CYPs. It is metabolized mainly to desethylchloroquine (DCQ), and, to a lesser extent, bisdesethyl-chloroquine. Using human liver microsomes, characterized for different isoforms, and recombinant enzymes, CYP2C8, CYP3A4/5, and CYP2D6 have been identified as the main isoforms responsible for CQ metabolism (Ducharme and Farinotti, 1996; Li *et al.*, 2003). Contradictory results have been presented regarding the inhibition of CYP2D6 activity by CQ in healthy human volunteers: Simooya *et al.* (1998) and Adedoyin *et al.* (1998) have shown that CQ causes a significant decrease in the activity, while Masimirembwa *et al.* (1996) reported that CQ has no significant effect, although there was a trend towards a decrease in the activity.

Amodiaquine metabolism to N-desethylamodiaquine is the principal route of disposition in humans (Li *et al.*, 2002). Using human liver microsomes and recombinant enzymes, CYP2C8 was identified as the main hepatic isoform that metabolized amodiaquine. The extrahepatic CYP1A1 and CYP1B1 also showed catalytic activity for the drug (Li *et al.*, 2002, 2003).

Mefloquine and primaquine are metabolized to carboxymefloquine and carboxy-primaquine, respectively, by human liver microsomes (Bangchang *et al.*, 1992a, 1992b). In human liver cells and microsomes, another metabolite of mefloquine, hydroxylmefloquine, was also detected (Fontaine *et al.*, 2000). The involvement of CYP3A4 in mefloquine biotransformation was suggested by several lines of evidence (Baune *et al.*, 1999; Fontaine *et al.*, 2000), while CYP1A2 and CYP3A4 were identified as the hepatic enzymes responsible for primaquine metabolism (Li *et al.*, 2003).

In humans, halofantrine and lumefantrine are metabolized to desbutyl derivatives. In human liver microsomes, both CYP3A4 and CYP3A5 were found to metabolize halofantrine, with major involvement of CYP3A4 (Baune *et al.*, 1999). Although halofantrine is a potent inhibitor of CYP2D6 activity (Simooya *et al.*, 1998), this enzyme did not mediate the metabolism of halofantrine (Lefevre *et al.*, 2000). Lumefantrine is also predominantly metabolized through CYP3A4 (Lefevre *et al.*, 2000; Giao and de Vries, 2001).

5.3.1. *Metabolism of Antifolate Combination Drugs*

Dapsone is metabolized to hydroxylamine-dapsone and monoacetyl-dapsone (Zuidema *et al.*, 1986). In human liver microsomes, CYP2C9 (Winter *et al.*, 2000) as well as CYP3A4 (Li *et al.*, 2003) predominantly contribute to N-hydroxylation. *In vivo*, CYP3A4 has been identified as an important mediator of this reaction (May *et al.*, 1994).

Proguanil is metabolized to cycloguanil. *In vitro*, proguanil activation to cycloguanil is mediated by CYP2C19 (~73%) and CYP3A4 (16%) (Birkett *et al.*, 1994; Lu *et al.*, 2000). *In vivo*, CYP2C19 is the major isoform responsible for the formation of cycloguanil (Ward *et al.*, 1991; Helsby *et al.*, 1993; Somogyi *et al.*, 1996).

Apart from dapsone and proguanil, not much is known about the metabolism of other antimalarial sulfa drugs and DHFR inhibitors. The CYP-mediated degradation is not a major metabolic pathway of sulfonamides. However, there is evidence that sulfamethoxazole may inhibit CYP2C9 activity (Miners and Birkett, 1998). Similarly, trimethoprim, a DHFR inhibitor, may inhibit CYP2C8/9 activity (Giao and de Vries, 2001; Wen *et al.*, 2002). Using recombinant enzymes at relatively high concentrations, it was observed that CYP1B1 and CYP2C19 contributed 11 and 13%, respectively, to the metabolism of pyrimethamine (Li *et al.*, 2003).

Atovaquone, used in combination with proguanil (Malarone), has been found to inhibit CYP3A4 activity of the recombinant enzyme (Thapar *et al.*, 2002) and CYP2C9 activity in human liver microsomes (Miller and Trepanier, 2002). However, the role of these enzymes in the metabolism of this drug is not clear. There is indirect evidence that atovaquone may undergo limited metabolism; however, a specific metabolite has not been identified.

5.3.2. *Involvement of CYPs and UGTs in ART Drug Metabolism*

Artemisinin and its derivatives are metabolized in the liver, mainly via Phase I oxidative dealkylation reaction (Navaratnam *et al.*, 2000) (Brossi A, cited in WHO/TDR publication #63, 2004) (Table 5.5).

Using the human liver microsomes, it was found that *in vitro* metabolism of artemisinin was mediated primarily by CYP2B6, with probable secondary contributions of CYP3A4 and CYP2A6 (Svensson and Ashton, 1999). Negligible metabolism was observed with recombi-

Table 5.5. Artemisinin Drugs and Their Metabolism

Compound	Isoform	Major metabolite	Chromosome
Artemisinin	CYP2B6, CYP3A4, CYP2A6	?	19, 7, 19
Dihydroartemisinin	UGT1A9, UGT2B7	DHA-glucuronide	2, 4
Artesunate	CYP2A6, CYP2B6	DHA	19
Artemether	CYP3A4*	DHA	7
Arteether	CYP3A4/3A5, CYP2B6	DHA	7, 19
Artelinic acid	CYP3A4/3A5	DHA	7

?: Not well characterized.
* Suggested but not conclusive.

nant enzymes CYP1A1, 1A2, 2C8, 2C9, 2C19, 2D6, and 2E1. In another *in vitro* study, the involvement of CYP2B6 and CYP3A4 in the metabolism of artemisinin was confirmed (Li *et al.*, 2003). Recently, the primary involvement of CYP2B6 in the metabolism of artemisinin was established *in vivo* in healthy human subjects (Simonsson *et al.*, 2003). Consistent with the *in vitro* results mentioned above, CYP2C9 and CYP2C19 did not seem to play a role in artemisinin metabolism *in vivo* (Simonsson *et al.*, 2003). The *in vitro* study by Li *et al.* (2003) also identified CYP2A6 to be primarily responsible for artesunate metabolism, with minor involvement of CYP2B6.

In the case of artemether metabolism, no major contribution of the enzymes CYP2D6 or CYP2C19 was found (van Agtmael *et al.*, 1998), and the role of CYP3A4, although suggested, is not clear (Lefevre *et al.*, 2000; van Agtmael *et al.*, 1999a, 1999b). However, *in vitro* study provided strong evidence that CYP3A4 is the primary enzyme involved in beta-arteether metabolism, with secondary contributions of CYP2B6 and CYP3A5 (Grace *et al.*, 1998). The enzymes CYP3A4 and CYP3A5 have been shown to be involved in the metabolism of artelinic acid, a water-soluble ART analogue (Grace *et al.*, 1999).

Recent *in vitro* studies revealed that CYP-catalyzed oxidation did not play a significant role in the metabolism of dihydroartemisinin, which is the major metabolite of ART derivatives (Bapiro *et al.*, 2002b; Ilett *et al.*, 2002). Dihydroartemisinin is converted to dihydroartemisinin–glucuronide by a glucuronidation reaction (O'Neill *et al.*, 2001), catalyzed by Phase II drug metabolizing enzymes UGT1A9 and UGT2B7 (Ilett *et al.*, 2002).

The ART drugs, in combination with other antimalarials, are viewed as the most promising drug regimen in the treatment of falciparum malaria (Olliaro and Taylor, 2004). Therefore, it is important to identify the factors that may increase the risk of adverse consequences of drug–drug interactions. Understanding the role of P450s in drug metabolism and the effects of drugs on P450s, the likelihood of adverse drug–drug interactions can be minimized. For this, two sets of information are needed: (a) to what extent is a specific P450 enzyme responsible for the metabolism of specific drugs and (b) to what extent will a given drug induce or inhibit a specific P450 enzyme. With these two data sets, one can deduce whether the addition of a second drug to an ongoing regimen is likely to cause a clinically meaningful change in the efficacy of the first drug. Artemisinin may induce CYP2C19 (Svensson *et al.*, 1998; Mihara *et al.*, 1999), but have no effect on CYP3A4 activity (Svensson *et al.*, 1998). This phenomenon may have a significant clinical implication in artemisinin-based combination therapy, as CYP3A4 is frequently involved in the metabolism of other antimalarial agents, such as mefloquine and lumefantrine. The combination of ART drugs with each other, with mefloquine, and the artemether–lumefantrine combination have been used in various clinical studies, and no significant drug interactions were observed (Giao and de Vries, 2001). Significant positive correlations and synergism were observed *in vitro* for various combinations of ART derivatives with other antimalarial drugs in several studies (Bustos *et al.*, 1994; Ringwald *et al.*, 1999; Gupta *et al.*, 2002a, 2002b).

5.4. Antimalarial Drug Levels and Treatment Outcome

Although the previous subsection identifies involvement of CYP and UGT enzyme systems in the metabolism of various antimalarials, whether there is any evidence that antimalarial drug levels vary and whether this variation affects the treatment outcome.

5.4.1. *In vivo CQ Levels and Parasite Drug Response Phenotypes*

An earlier study determined the concentrations of CQ and DCQ in African children with acute falciparum malaria (Walker *et al.*, 1983). Peak plasma CQ and DCQ concentrations ranged between 65 and 263 ng/ml and 9 and 62 ng/ml, respectively. Despite this variability, effective clinical and parasitological response to treatment was observed within 2–4 days in all patients. Nevertheless, Walker *et al.* (1983) indicated the possibility that the unchanged drug and its metabolite concentrations in plasma could play a role in the rate of development of CQR in different geographic areas. In Tanzanian schoolchildren, pronounced interindividual variability in CQ (3.3- to 5.1-fold) and DCQ (3.5- to 6.3-fold) concentrations in whole blood during and after treatment was observed

(Hellgren *et al.*, 1989). In subjects, in which parasites were detected on day 7 after initiation of the treatment (=RII response), the mean highest CQ concentration was significantly lower than that in subjects without parasites (1473 versus 1799 nmol/L). In a review of CQ treatment/prophylaxis studies, an interindividual variation of 2.5- to 5.6-fold was found (Hellgren *et al.*, 1993). In a later study (Hellgren *et al.*, 1994), The CQ and DCQ concentrations in whole blood of Tanzanian schoolchildren were found to vary between 6 and 950 nmol/L and 10 and 299 nmol/L, respectively. Only 9% of the children had CQ concentrations >100 nmol/L (i.e., curative level). In these children, *P. falciparum* trophozoites were significantly less common compared with the others. Hellgren *et al.* (1994) suggested that small, subtherapeutic doses of CQ taken frequently could promote development of resistance. A number of other studies have noted that young children are more likely to fail CQ treatment than older malaria patients due to lower plasma CQ concentrations (Maitland *et al.*, 1997; Mockenhaupt *et al.*, 2000; Ringwald *et al.*, 2000). Maitland *et al.* (1997) and Mockenhaupt *et al.* (2000) also suggested that subtherapeutic plasma drug levels were likely to further promote CQR, most likely by constituting a predisposing environment for the selection and spread of resistant *P. falciparum* strains (Wernsdorfer, 1992).

In some studies, the plasma levels of CQ and DCQ in patients infected with sensitive and resistant malaria parasites were compared. Large interindividual variations (20–35%) were observed on day 3 in the mean plasma CQ concentrations of patients from Sudan (Karim *et al.*, 1992). The mean plasma CQ level in the sensitive group was higher than that in the RIII resistant group (0.275 versus 0.225 µmol/L). Similarly, the mean plasma CQ:DCQ ratio in the sensitive group was higher than that in the RIII resistant group (3.14 versus 3.05). However, the differences in the mean plasma CQ level and CQ:DCQ ratio were not significant. In another study, concentrations of CQ and DCQ in blood cells and plasma from CQS and CQR cases from India were determined on day 2 and day 7 after initiation of treatment (Dua *et al.*, 2000). On day 2, the mean CQ concentrations in the sensitive cases were significantly higher than those in the resistant patients [both in plasma (0.47 versus 0.32 µg/ml) and in the blood cells (1.51 versus 0.46 µg/ml)], and the mean CQ:DCQ ratio was significantly higher in the blood cells from the sensitive group than in the blood cells from the resistant cases (3.07 versus 1.77). With these results in mind, results of May and Meyer (2003) are intriguing. They found that the prevalence of CQR-associated *pfcrt* T76 variant increased substantially with the CQ blood concentrations measured in asymptomatic *P. falciparum*-infected children from Nigeria. Blood levels of CQ were significantly higher in lower age group children (<2 to 3 years old) compared with higher age group children (4 to >6 years old). Only the *pfcrt* T76 variant was observed in younger children with CQ levels >150 nmol/L, whereas CQS-associated K76 allele was more frequent (~80%) in older children without detectable plasma CQ levels. Based on these observations, the authors suggested that the prevalence of *pfcrt* T76 is a function of the actual CQ level and, thus, of age, and possibly influenced by acquired immunity and natural resistance factors of the host.

Overall, the studies described above suggest that (1) the blood concentrations of an antimalarial drug can differ in different age group individuals, (2) large interindividual variability can occur in the blood concentrations of a drug/metabolite, and (3) variability in the blood concentrations of a drug/metabolite may affect the antimalarial treatment efficacy, and may contribute to the selection of resistant parasites. Lower blood CQ concentrations could be attributable to patient noncompliance with drug intake. Young children might have difficulty in ingesting and retaining all of the prescribed medication, which could contribute to their lower blood CQ concentrations. However, in the studies (Walker *et al.*, 1983; Hellgren *et al.*, 1989, 1993; Ringwald *et al.*, 2000; Dua *et al.*, 2000) patients received the treatment under supervision. Therefore, lower CQ levels observed in these studies are probably due to variability in the metabolism of the drug.

5.4.2. *Variability in the Metabolism of Proguanil*

Metabolism of the antimalarial drug proguanil and the anticonvulsant drug mephenytoin is mediated via CYP2C19 (Helsby *et al.*, 1993). The PM, IM, and EM phenotypes are identified by measuring the proguanil:cycloguanil metabolic ratio in the urine or plasma, the ability to metabolize *S*-mephenytoin, and/or by genotype analysis. The prevalence of these drug metabolism phenotypes varies substantially in populations with different ethnic backgrounds (Yusuf *et al.*, 2003), and two defective alleles, *CYP2C19*2* and *CYP2C19*3*, are largely associated with the PM phenotype (De Morais *et al.*, 1994; Persson *et al.*, 1996). The frequencies of the PM phenotype in various populations are provided in Table 5.6 (Watkins *et al.*, 1990; Masimirembwa *et al.*, 1995; Wanwimolruk *et al.*, 1995; Skjelbo *et al.*, 1996; Kaneko *et al.*, 1997; Hoskins *et al.*, 1998; Wanwimolruk *et al.*, 1998; Goldstein, 2001; Griese *et al.*, 2001; Bolaji *et al.*, 2002; Masta *et al.*, 2003).

An earlier study found a correlation between the *in vivo* proguanil:cycloguanil ratio and the number of malaria breakthrough par-

Table 5.6. CYP2C19 Poor Metabolism (PM) Phenotype Frequencies in Various Populations

Population	Frequency (%)
Kenyan	35
Zimbabwean and Tanzanian, Ethiopian and Nigerian	4–5.2
Asian	12–23
Caucasian	2–5
Maori (New Zealand)	7
Australian Aboriginal	25.6
South Pacific Polynesian	13.6
Vanuatuan	70.6
Sepik (Papua New Guinea)	36

asitemia episodes (Skjelbo *et al.*, 1996). This study suggested that the impaired cycloguanil formation might be a contributing risk factor in malaria prophylaxis with proguanil. However, the causal prophylactic effect of proguanil appeared to be unrelated to the metabolic phenotype (either PM or EM) in Japanese volunteers in Kenya (Mberu *et al.*, 1995), suggesting that poor metabolism of proguanil, and the resulting low plasma concentrations of cycloguanil, do not lead to an increased breakthrough frequency of proguanil chemoprophylaxis.

Plasma samples, obtained from PM and EM Thai patients treated with proguanil-atovaquone, were assessed for the *in vitro* antimalarial activity (Edstein *et al.*, 1996). The activity was similar in both the phenotype groups. These findings suggested that the phenotypic status of Thai patients did not alter the *in vitro* antimalarial activity of proguanil combined with atovaquone (Edstein *et al.*, 1996), although this may be explained by the specific synergism between the two drugs (Edstein *et al.*, 1996; Srivastava *et al.*, 1999). In a later study, therapeutic efficacy of proguanil treatment in PM patients was found to be similar to that in EM patients (Kaneko *et al.*, 1999). There was even a trend toward lower efficacy in EM patients (the mean therapeutic efficacy for falciparum and vivax infection, PM = 75 and 91%, EM = 64 and 88%, respectively). The authors suggested that the parent compound proguanil may be more active than its main metabolite cycloguanil, and that it may have a significant intrinsic efficacy independent of cycloguanil. Another plausible explanation might be that an as yet-undefined metabolite through another metabolic pathway is responsible for the proguanil efficacy in PMs. Therefore, further studies are needed to establish the clinical significance of CYP2C19 genotype in the antimalarial efficacy of proguanil.

5.4.3. Variability in the Pharmacokinetics of ART Drugs

The ART drugs are available for oral, rectal (suppository), and IM/IV administrations (Navaratnam *et al.*, 2000; White, 2003). ART drugs are rapidly absorbed and eliminated. Pharmacokinetic parameters of these drugs are highly variable. Time to peak plasma levels for ART drugs varies from minutes to hours, depending on the drug formulation and its route of administration (White, 2003; de Vries and Dien, 1996). Bioavailability of ART drugs is also highly variable (<25% to >85%), depending on the drug formulation, its route of administration, health status of the individual, and the nature of malaria infection (Navaratnam *et al.*, 2000). All ART drugs, except artemisinin, are hydrolyzed to the active metabolite dihydroartemisinin (Lee and Hufford, 1990; Zhang *et al.*, 2001). Biotransformation into the active metabolite dihydroartemisinin occurs almost immediately (<15 min) for artesunate. Dihydroartemisinin has an elimination half-life of approximately 45 min (White, 2003). The elimination half-life of artemisinin is 2–5 h (de Vries and Dien, 1996; White, 1994). Evidence suggests rapid elimination of artesunate (in minutes) and artemether (1–11 h) as well (de Vries and Dien, 1996; White, 2003).

Large interindividual variations in pharmacokinetic parameters of artemisinin, including the maximum concentration in blood, have been observed in healthy subjects after oral administration (range 299–888 ng/ml) (Dien *et al.*, 1997), and in malaria patients after rectal suppository administration (range 29–169 ng/ml) (Koopmans *et al.*, 1998). Interindividual variability was higher in healthy subjects after rectal administration (range 24–330 ng/ml) than after an oral dosage of the drug (Koopmans *et al.*, 1998, 1999). Substantial interindividual variations in artesunate/dihydroartemisinin pharmacokinetic parameters were also noted after rectal administration of artesunate in children with uncomplicated or moderate malaria in Gabon (Halpaap *et al.*, 1998), Ghana (Krishna *et al.*, 2001), and PNG (Karunajeewa *et al.*, 2004). In Thailand, adult patients with acute uncomplicated malaria were given a single oral dose of artesunate varying between 25 and 250 mg (Angus *et al.*, 2002). In these patients, dihydroartemisinin concentrations at 2 h varied between 2 and 1212-fold (range 0–2155 ng/ml), depending on the dose of artesunate given.

As discussed earlier, CYPs are involved in the metabolism of artemisinin and its most commonly used derivatives—artesunate and artemether. Similarly, UGTs are involved in the metabolism of dihydroartemisinin, the active metabolite of artesunate and artemether. Although both CYPs and UGTs are polymorphic, it is not known whether these polymorphisms affect the metabolism and, consequently, plasma levels of ART drugs. It is estimated that the lowest effective plasma concentration is in the range 3–30 ng/ml (Koopmans *et al.*, 1998), which corresponds to IC_{50} values 1–40 ng/ml *in vitro* (ter Kuile *et al.*, 1993). However, a minimum parasiticidal concentration *in vivo* is not yet defined. In spite of large interindividual variations in the peak plasma concentration, bioavailability, and other pharmacokinetic parameters, it is claimed that ART drugs were effective in parasite and fever clearance in patients with uncomplicated or moderate malaria (Koopmans *et al.*, 1998; Halpaap *et al.*, 1998; Krishna *et al.*, 2001; Angus *et al.*, 2002). This said, other antimalarial drugs, such as mefloquine (Koopmans *et al.*, 1998; Angus *et al.*, 2002), sulfadoxine–pyrimethamine (Halpaap *et al.*, 1998), or chloroquine or sulfadoxine–pyrimethamine (Krishna *et al.*, 2001), were also given to the patients. Therefore, one cannot conclude from these studies that the subtherapeutic plasma levels of ART drugs did not occur and, due to that, the ART drug efficacy was not compromised. It is possible that in the patients with subtherapeutic plasma levels of ART drugs, the parasite and fever clearance was due to the other antimalarial drug.

6. Conclusions

Malaria control has relied on (1) killing the mosquito vector and (2) an effective chemotherapy/chemoprophylaxis. An effective malaria vaccine is not yet available, although substantial, promising advancements have been made (Alonso *et al*, 2004). It is believed that the extensive deployment of antimalarial drugs, in the past 50 years, has provided a tremendous selection pressure on human malaria parasites to evolve mechanisms of resistance. The emergence of drug resistance, particularly in *P. falciparum*, has been a major contributor to the global resurgence of malaria in the last three decades. Resistance has already developed to all the antimalarial drug classes, and seems to be developing for the newer series of compounds—the artemisinins. These drugs are already an essential component of treatments for multidrug-resistant falciparum malaria. It is true that "if we lose artemisinins to resistance, we may be faced with untreatable malaria" (White, 2004).

Malaria is a blood infection. The response to treatment is determined by antimalarial drug concentrations in the blood, and variability in drug concentrations can contribute to the emergence of drug resistance (Hastings *et al.*, 2002). Variability in drug concentrations could be due to variability in (a) absorption and distribution of the drug or (b) metabolism and excretion of the drug. In the liver, CYP and UGT enzyme superfamilies are involved in the metabolism of drugs, including many of the antimalarials. Polymorphisms are observed in several members of these enzyme superfamilies. These polymorphisms are inherited, and, therefore, are a major cause of interindividual and interethnic variability in drug metabolism. The variability in drug metabolism is one of the important factors that may cause drug concentrations to fall outside of the therapeutic range, and, thus, may affect the efficacy of the drug.

Although polymorphisms in parasite genes play an important role in antimalarial drug resistance, the role of these human drug metabolizing enzyme – CYP and UGT – polymorphisms in the antimalarial drug resistance, and drug efficacy, has not been explored. In addition to the extensive and unregulated usage of antimalarial drugs, CYP and UGT polymorphisms can contribute to the selection of resistant parasites by altering drug/metabolite concentrations in the blood that are required to kill the parasites. As the exposure of malaria parasites to suboptimal antimalarial drug levels is an important factor contributing to the evolution of antimalarial drug resistance (White, 2004; Peters, 1987), polymorphisms in the human drug metabolizing proteins must be contributing to the selection of these parasites.

An integrated approach for assessing treatment response in falciparum malaria is urgently needed (Winstanley and Watkins, 1992; Cox-Singh *et al.*, 2003). This type of approach would consider chemosensitivity of given parasite strains as well as host metabolic variability in drug response. Such an approach would promote a rational antimalarial use, thereby preserving the therapeutic efficacy of promising and valuable drugs in regions where antimalarial treatment failure poses a continuing public health problem.

Acknowledgments

Our sincere thanks are due to Charles King, David McNamara, Brian Grimberg, and Jeana DaRe for critically reading the manuscript and helpful suggestions.

References

Abraham, B.K. and Adithan, C. (2001). Genetic polymorphism of *CY2D6*. *Indian J. Pharmacol.*, **33**, 147–169.

Adedoyin, A., Frye, R.F., Mauro, K. *et al.* (1998). Chloroquine modulation of specific metabolizing enzymes activities: investigation with selective five drugs cocktail. *Br. J. Clin. Pharmacol.*, **46**, 215–219.

Ahmed, A., Bararia, D., Vinayak, S. *et al.* (2004). *Plasmodium falciparum* isolates in India exhibits a progressive increase in mutations associated with sulfadoxine–pyrimethamine resistance. *Antimicrob. Agents Chemother.*, **48**, 879–889.

Alonso, P.L., Sacarlal, J., Aponte, J.J. *et al.* (2004). Efficacy of the RTS,S/AS02A vaccine against *Plasmodium falciparum* infection and disease in young African children: Randomized controlled trial. *Lancet*, **364,** 1411–1420.

Anderson, T.J. (2004). Mapping drug-resistance genes in *Plasmodium falciparum* by genome-wide association. *Curr. Drug Targets Infect. Disord.*, **4,** 65–78.

Anderson, T.J., Nair, S., Jacobzone, C. *et al.* (2003). Molecular assessment of drug resistance in *Plasmodium falciparum* from Bahr El Gazal province, Sudan. *Trop. Med. Int. Health*, **8,** 1068–1073.

Angus, B.J., Thaiaporn, I., Chanthapadith, K. *et al.* (2002). Oral artesunate dose–response relationship in acute falciparum malaria. *Antimicrob. Agents Chemother.*, **46,** 778–782.

Baird, J.K., Basri, H., Purnomo *et al.* (1991). Resistance to chloroquine by *Plasmodium vivax* in Irian Jaya, Indonesia. *Am. J. Trop. Med. Hyg.*, **44,** 547–552.

Baird, J.K., Sustriayu Nalim, M.F., Basri, H. *et al.* (1996). Survey of resistance to chloroquine by *Plasmodium vivax* in Indonesia. *Trans. R. Soc. Trop. Med. Hyg.*, **90,** 409–411.

Balint, G.A. (2001). Artemisinin and its derivatives: An important new class of antimalarial agents. *Pharmaco. Ther.*, **90,** 261–265.

Ball, S. and Borman, N. (1997). Pharmacogenetics and drug metabolism. *Nat. Biotechnol.*, **15,** 925–926.

Bangchang, K.N., Karbwang, J., and Back, D.J. (1992a). Mefloquine metabolism by human liver microsomes. Effect of other antimalarial drugs. *Biochem. Pharmacol.*, **43,** 1957–1961.

Bangchang, K.N., Karbwang, J., and Back, D.J. (1992b). Primaquine metabolism by human liver microsomes: effect of other antimalarial drugs. *Biochem. Pharmacol.*, **44,** 587–590.

Bapiro, T.E., Egnell, A.C., Hasler, J.A. *et al.* (2001). Application of higher throughput screening (HTS) inhibition assays to evaluate the interaction of antiparasitic drugs with cytochrome P450s. *Drug Metab. Dispos.*, **29,** 30–35.

Bapiro, T.E., Andersson, T.B., Otter, C. *et al.* (2002a). Cytochrome P450 1A1/2 induction by antiparasitic drugs: Dose-dependent increase in ethoxyresorufin O-deethylase activity and mRNA caused by quinine, primaquine and albendazole in HepG2 cells. *Eur. J. Clin. Pharmacol.*, **58,** 537–542.

Bapiro, T.E., Hasler, J.A., Ridderstrom, M. *et al.* (2002b). The molecular and enzyme kinetic basis for the diminished activity of the cytochrome P450 2D6.17 (CYP2D6.17) variant. Potential implications for CYP2D6 phenotyping studies and the clinical use of CYP2D6 substrate drugs in some African populations. *Biochem. Pharmacol.*, **64,** 1387–1398.

Basco, L.K., Same-Ekobo, A., Ngane, V.F. *et al.* (2002). Therapeutic efficacy of sulfadoxine–pyrimethamine, amodiaquine and the sulfadoxine–pyrimethamine-amodiaquine combination against uncomplicated *Plasmodium falciparum* malaria in young children in Cameroon. *Bull. World Health Organ.*, **80,** 538–545.

Baune, B., Flinois, J.P., Furlan, V. *et al.* (1999). Halofantrine metabolism in microsomes in man: major role of CYP 3A4 and CYP 3A5. *J. Pharm. Pharmacol.*, **51,** 419–426.

Baune, B., Furlan, V., Taburet, A.M. *et al.* (1999). Effect of selected antimalarial drugs and inhibitors of cytochrome P-450 3A4 on halofantrine metabolism by human liver microsomes. *Drug Metab. Dispos.*, **27,** 565–568.

Beaulieu, M., Levesque, E., Tchernof, A. *et al.* (1997). Chromosomal localization, structure, and regulation of the *UGT2B17* gene, encoding a C19 steroid metabolizing enzyme. *DNA Cell. Biol.*, **16,** 1143–1154.

Bennett, T.N., Kosar, A.D., Ursos, L.M. *et al.* (2004). Drug resistance-associated pfCRT mutations confer decreased *Plasmodium falciparum* digestive vacuolar pH. *Mol. Biochem. Parasitol.*, **133,** 99–114.

Birkett, D.J., Rees, D., Andersson, T. *et al.* (1994). *In vitro* proguanil activation to cycloguanil by human liver microsomes is mediated by CYP3A isoforms as well as by S-mephenytoin hydroxylase. *Br. J. Clin. Pharmacol.*, **37,** 413–420.

Bloland, P.B. (2001). Drug resistance in malaria. WHO/CDS/CSR/DRS/2001. **4,** 1–27.

Bloland, P.B., Ettling, M., and Meek, S. (2000). Combination therapy for malaria in Africa: hype or hope? *Bull. World Health Organ.*, **78,** 1378–1388.

Bock, K.W. (2003). Vertebrate UDP-glucuronosyltransferases: Functional and evolutionary aspects. *Biochem. Pharmacol.*, **66,** 691–696.

Bolaji, O.O., Sadare, I.O., Babalola, C.P. *et al.* (2002). Polymorphic oxidative metabolism of proguanil in a Nigerian population. *Eur. J. Clin. Pharmacol.*, **58,** 543–545.

Brabin, B.J., Ginny, M., Sapau, J. *et al.* (1990). Consequences of maternal anaemia on outcome of pregnancy in a malaria endemic area in Papua New Guinea. *Ann. Trop. Med. Parasitol.*, **84,** 11–24.

Brega, S., Meslin, B., de Monbrison, F. *et al.* (2005). Identification of the *Plasmodium vivax* mdr-like gene (*pvmdr1*) and analysis of single-nucleotide polymorphisms among isolates from different areas of endemicity. *J. Infect. Dis.*, **191,** 272–277.

Brockman, A., Price, R.N., van Vugt, M. *et al.* (2000). *Plasmodium falciparum* antimalarial drug susceptibility on the northwestern border of Thailand during five years of extensive use of artesunate–mefloquine. *Trans. R. Soc. Trop. Med. Hyg.*, **94,** 537–544.

Brooks, D.R., Wang, P., Read, M. *et al.* (1994). Sequence variation of the hydroxymethyldihydropterin pyrophosphokinase: Dihydropteroate synthase gene in lines of the human malaria parasite, *Plasmodium falciparum*, with differing resistance to sulfadoxine. *Eur. J. Biochem.*, **224,** 397–405.

Brossi, A., Venugopalan, B., Dominguez Gerpe, L. *et al.* (1988). Arteether, a new antimalarial drug: synthesis and antimalarial properties. *J. Med. Chem.*, **31,** 645–650.

Burchell, B. (2003). Genetic variation of human UDP-glucuronosyltransferase: Implications in disease and drug glucuronidation. *Am. J. Pharmacogenomics*, **3,** 37–52.

Bustos, M.D., Gay, F., and Diquet, B. (1994). *In-vitro* tests on Philippine isolates of *Plasmodium falciparum* against four standard antimalarials and four qinghaosu derivatives. *Bull. World Health Organ.*, **72,** 729–735.

Bwijo, B., Kaneko, A., Takechi, M. *et al.* (2003). High prevalence of quintuple mutant *dhps/dhfr* genes in *Plasmodium falciparum* infections seven years after introduction of sulfadoxine and pyrimethamine as first line treatment in Malawi. *Acta. Trop.*, **85,** 363–373.

Bzik, D.J., Li, W.B., Horii, T. *et al.* (1987). Molecular cloning and sequence analysis of the *Plasmodium falciparum* dihydrofolate reductase–thymidylate synthase gene. *Proc. Natl. Acad. Sci. USA*, **84,** 8360–8364.

Caraco, Y. (1998). Genetic determinants of drug responsiveness and drug interactions. *Ther. Drug Monit.*, **20,** 517–524.

Carlton, J.M., Fidock, D.A., Djimde, A. *et al.* (2001). Conservation of a novel vacuolar transporter in *Plasmodium* species and its central role in chloroquine resistance of *Plasmodium falciparum*. *Curr. Opin. Microbiol.*, **4,** 415–420.

Cerutti, C. Jr., Durlacher, R.R., de Alencar, F.E. *et al.* (1999). *In vivo* efficacy of mefloquine for the treatment of Falciparum malaria in Brazil. *J. Infect. Dis.*, **180,** 2077–2080.

Chen, G.X., Mueller, C., Wendlinger, M. *et al.* (1987). Kinetic and molecular properties of the dihydrofolate reductase from pyrimethamine-sensitive and pyrimethamine-resistant clones of the human malaria parasite *Plasmodium falciparum*. *Mol. Pharmacol.*, **31,** 430–437.

Chen, N., Russell, B., Fowler, E. *et al.* (2002). Levels of chloroquine resistance in *Plasmodium falciparum* are determined by loci other than *pfcrt* and *pfmdr1*. *J. Infect. Dis.*, **185,** 405–407.

Chotivanich, K., Udomsangpetch, R., Chierakul, W. *et al.* (2004). *In vitro* efficacy of antimalarial drugs against *Plasmodium vivax* on the western border of Thailand. *Am. J. Trop. Med. Hyg.*, **70,** 395–397.

Collins, W.E. and Jeffery, G.M. (1996). Primaquine resistance in *Plasmodium vivax*. *Am. J. Trop. Med. Hyg.*, **55,** 243–249.

Collins, W.E., Jeffery, G.M., and Roberts, J.M. (2003). A retrospective examination of anemia during infection of humans with *Plasmodium vivax*. *Am. J. Trop. Med. Hyg.*, **68,** 410–412.

Contreras, C.E., Cortese, J.F., Caraballo, A. *et al.* (2002). Genetics of drug-resistant *Plasmodium falciparum* malaria in the Venezuelan state of Bolivar. *Am. J. Trop. Med. Hyg.*, **67,** 400–405.

Cooper, R.A., Ferdig, M.T., Su, X.Z. *et al.* (2002). Alternative mutations at position 76 of the vacuolar transmembrane protein PfCRT are associated with chloroquine resistance and unique stereospecific quinine and quinidine responses in *Plasmodium falciparum*. *Mol. Pharmacol.*, **61,** 35–42.

Cortese, J.F., Caraballo, A., Contreras, C.E. *et al.* (2002). Origin and dissemination of *Plasmodium falciparum* drug resistance mutations in South America. *J. Infect. Dis.*, **186,** 999–1006.

Cowman, A.F., Galatis, D., and Thompson, J.K. (1994). Selection for mefloquine resistance in *Plasmodium falciparum* is linked to amplification of the *pfmdr1* gene and cross-resistance to halofantrine and quinine. *Proc. Natl. Acad. Sci. USA*, **91,** 1143–1147.

Cowman, A.F., Karcz, S., Galatis, D. *et al.* (1991). A P-glycoprotein homologue of *Plasmodium falciparum* is localized on the digestive vacuole. *J. Cell Biol.*, **113,** 1033–1042.

Cowman, A.F., Morry, M.J., Biggs, B.A. *et al.* (1988). Amino acid changes linked to pyrimethamine resistance in the dihydrofolate reductase–thymidylate synthase gene of *Plasmodium falciparum. Proc. Natl. Acad. Sci. USA*, **85,** 9109–9113.

Cox-Singh, J., Lu, H.Y., Davis, T.M. *et al.* (2003). Application of a multifaceted approach for the assessment of treatment response in falciparum malaria: A study from Malaysian Borneo. *Int. J. Parasitol.*, **33,** 1545–1552.

De Morais, S.M., Wilkinson, G.R., Blaisdell, J. *et al.* (1994). Identification of a new genetic defect responsible for the polymorphism of (S)-mephenytoin metabolism in Japanese. *Mol. Pharmacol.*, **46,** 594–598.

de Pecoulas, P.E., Basco, L.K., Tahar, R. *et al.* (1998a). Analysis of the *Plasmodium vivax* dihydrofolate reductase–thymidylate synthase gene sequence. *Gene*, **211,** 177–185.

de Pecoulas, P.E., Tahar, R., Ouatas, T. *et al.* (1998b). Sequence variations in the *Plasmodium vivax* dihydrofolate reductase–thymidylate synthase gene and their relationship with pyrimethamine resistance. *Mol. Biochem. Parasitol.*, **92,** 265–273.

de Vries, P.J. and Dien, T.K. (1996). Clinical pharmacology and therapeutic potential of artemisinin and its derivatives in the treatment of malaria. *Drugs*, **52,** 818–836.

Demar, M. and Carme, B. (2004). *Plasmodium falciparum in vivo* resistance to quinine: Description of two RIII responses in French Guiana. *Am. J. Trop. Med. Hyg.*, **70,** 125–127.

Desowitz, R.S. and Spark, R.A. (1987). Malaria in the Maprik area of the Sepik region, Papua New Guinea: 1957–1984. *Trans. R. Soc. Trop. Med. Hyg.*, **81,** 175–176.

Destenaves, B. and Thomas, F. (2000). New advances in pharmacogenomics. *Curr. Opin. Chem. Biol.*, **4,** 440–444.

Dieckmann, A. and Jung, A. (1986). Mechanisms of sulfadoxine resistance in *Plasmodium falciparum. Mol. Biochem. Parasitol.*, **19,** 143–147.

Dien, T.K., de Vries, P.J., Khanh, N.X. *et al.* (1997). Effect of food intake on pharmacokinetics of oral artemisinin in healthy Vietnamese subjects. *Antimicrob. Agents Chemother.*, **41,** 1069–1072.

Djimde, A.A., Doumbo, O.K., Traore, O. *et al.* (2003). Clearance of drug-resistant parasites as a model for protective immunity in *Plasmodium falciparum* malaria. *Am. J. Trop. Med. Hyg.*, **69,** 558–563.

Dua, V.K., Gupta, N.C., Kar, P.K. *et al.* (2000). Chloroquine and desethylchloroquine concentrations in blood cells and plasma from Indian patients infected with sensitive or resistant *Plasmodium falciparum. Ann. Trop. Med. Parasitol.*, **94,** 565–570.

Dua, V.K., Kar, P.K., and Sharma, V.P. (1996). Chloroquine resistant *Plasmodium vivax* malaria in India. *Trop. Med. Int. Health*, **1,** 816–819.

Ducharme, J. and Farinotti, R. (1996). Clinical pharmacokinetics and metabolism of chloroquine. Focus on recent advancements. *Clin. Pharmacokinet.*, **31** 257–274.

Duraisingh, M.T., Roper, C., Walliker, D. *et al.* (2000). Increased sensitivity to the antimalarials mefloquine and artemisinin is conferred by mutations in the *pfmdr1* gene of *Plasmodium falciparum. Mol. Microbiol.*, **36,** 955–961.

Durand, R., Jafari, S., Vauzelle, J. *et al.* (2001). Analysis of *pfcrt* point mutations and chloroquine susceptibility in isolates of *Plasmodium falciparum. Mol. Biochem. Parasitol.*, **114,** 95–102.

Edstein, M.D., Yeo, A.E., Kyle, D.E. *et al.* (1996). Proguanil polymorphism does not affect the antimalarial activity of proguanil combined with atovaquone *in vitro. Trans. R. Soc. Trop. Med. Hyg.*, **90,** 418–421.

Evans, W.E. and Relling, M.V. (1999). Pharmacogenomics: Translating functional genomics into rational therapeutics. *Science*, **286,** 487–491.

Farnert, A., Lindberg, J., Gil, P. *et al.* (2003). Evidence of *Plasmodium falciparum* malaria resistant to atovaquone and proguanil hydrochloride: case reports. *BMJ.*, **326,** 628–629.

Faye, F.B., Spiegel, A., Tall, A. *et al.* (2002). Diagnostic criteria and risk factors for *Plasmodium ovale* malaria. *J. Infect. Dis.*, **186,** 690–695.

Ferdig, M.T., Cooper, R.A., Mu, J., *et al.* (2004). Dissecting the loci of low-level quinine resistance in malaria parasites. *Mol. Microbiol.*, **52,** 985–997.

Fidock, D.A., Nomura, T., Talley, A.K. *et al.* (2000). Mutations in the *P. falciparum* digestive vacuole transmembrane protein PfCRT and evidence for their role in chloroquine resistance. *Mol, Cell,* **6,** 861–871.

Fivelman, Q.L., Butcher, G.A., Adagu, I.S. *et al.* (2002). Malarone treatment failure and *in vitro* confirmation of resistance of *Plasmodium falciparum* isolate from Lagos, Nigeria. *Malar. J.*, **1,** 1.

Fontaine, F., de Sousa, G., Burcham, P.C. *et al.* (2000). Role of cytochrome P450 3A in the metabolism of mefloquine in human and animal hepatocytes. *Life Sci.*, **66,** 2193–2212.

Foote, S.J., Galatis, D., and Cowman, A.F. (1990). Amino acids in the dihydrofolate reductase–thymidylate synthase gene of *Plasmodium falciparum* involved in cycloguanil resistance differ from those involved in pyrimethamine resistance. *Proc. Natl. Acad. Sci. USA*, **87,** 3014–3017.

Foote, S.J., Kyle, D.E., Martin, R.K. *et al.* (1990). Several alleles of the multidrug-resistance gene are closely linked to chloroquine resistance in *Plasmodium falciparum. Nature*, **345,** 255–258.

Garavelli, P.L. and Corti, E. (1992). Chloroquine resistance in *Plasmodium vivax*: The first case in Brazil. *Trans. R. Soc. Trop. Med. Hyg.*, **86,** 128.

Gay, F., Ciceron, L., Litaudon, M. *et al.* (1994). *In-vitro* resistance of *Plasmodium falciparum* to qinghaosu derivatives in west Africa. *Lancet*, **343,** 850–851.

Giao, P.T. and de Vries, P.J. (2001). Pharmacokinetic interactions of antimalarial agents. *Clin. Pharmacokinet.*, **40,** 343–373.

Giao, P.T., Binh, T.Q., Kager, P.A. *et al.* (2001). Artemisinin for treatment of uncomplicated falciparum malaria: Is there a place for monotherapy? *Am. J. Trop. Med. Hyg.*, **65,** 690–695.

Giao, P.T., de Vries, P.J., Hung le, Q. *et al.* (2004). CV8, a new combination of dihydroartemisinin, piperaquine, trimethoprim and primaquine, compared with atovaquone–proguanil against falciparum malaria in Vietnam. *Trop. Med. Int. Health*, **9,** 209–216.

Goldenberg, R.L. and Thompson, C. (2003). The infectious origins of stillbirth. *Am. J. Obstet. Gynecol.*, **189,** 861–873.

Goldstein, J.A. (2001). Clinical relevance of genetic polymorphisms in the human CYP2C subfamily. *Br. J. Clin. Pharmacol.*, **52,** 349–355.

Gong, Q.H., Cho, J.W., Huang, T. *et al.* (2001). Thirteen UDPglucuronosyltransferase genes are encoded at the human *UGT1* gene complex locus. *Pharmacogenetics*, **11,** 357–368.

Gonzalez, I.J., Varela, R.E., Murillo, C. *et al.* (2003). Polymorphisms in *cg2* and *pfcrt* genes and resistance to chloroquine and other antimalarials *in vitro* in *Plasmodium falciparum* isolates from Colombia. *Trans. R. Soc. Trop. Med. Hyg.*, **97,** 318–324.

Goodman & Gilman's *The Pharmacological Basis of Therapeutics* (2001). J.G. Hardman, L.E. Limbird, and A.G. Gilman (eds), 10th ed. McGraw-Hill.

Grace, J.M., Aguilar, A.J., Trotman, K.M. *et al.* (1998). Metabolism of beta-arteether to dihydroqinghaosu by human liver microsomes and recombinant cytochrome P450. *Drug Metab. Dispos.*, **26,** 313–317.

Grace, J.M., Skanchy, D.J., and Aguilar, A.J. (1999). Metabolism of artelinic acid to dihydroqinqhaosu by human liver cytochrome P4503A. *Xenobiotica*, **29,** 703–717.

Griese, E.U., Ilett, K.F., Kitteringham, N.R. *et al.* (2001). Allele and genotype frequencies of polymorphic cytochromes P4502D6, 2C19, and 2E1 in aborigines from western Australia. *Pharmacogenetics*, **11,** 69–76.

Guengerich, F.P. (2003). Cytochromes P450, drugs, and diseases. *Mol. Interv.*, **3,** 194–204.

Guillemette, C. (2003). Pharmacogenomics of human UDP-glucuronosyltransferase enzymes. *Pharmacogenomics J.*, **3,** 136–158.

Gupta, S., Thapar, M.M., Mariga, S.T. *et al.* (2002a). *Plasmodium falciparum*: *in vitro* interactions of artemisinin with amodiaquine, pyronaridine, and chloroquine. *Exp. Parasitol.*, **100,** 28–35.

Gupta, S., Thapar, M.M., Wernsdorfer, W.H. *et al.* (2002b). *In vitro* interactions of artemisinin with atovaquone, quinine, and mefloquine against *Plasmodium falciparum. Antimicrob. Agents Chemother*., **46,** 1510–1515.

Guyatt, H.L. and Snow, R.W. (2001). Malaria in pregnancy as an indirect cause of infant mortality in sub-Saharan Africa. *Trans. R. Soc. Trop. Med. Hyg*., **95,** 569–576.

Halpaap, B., Ndjave, M., Paris, M. *et al.* (1998). Plasma levels of artesunate and dihydroartemisinin in children with *Plasmodium falciparum* malaria in Gabon after administration of 50-mg artesunate suppositories. *Am. J. Trop. Med. Hyg*., **58,** 365–368.

Hartl, D.L. (2004). The origin of malaria: mixed messages from genetic diversity. *Nat. Rev. Microbiol*., **2,** 15–22.

Hastings, I.M., Watkins, W.M., and White, N.J. (2002). The evolution of drug-resistant malaria: The role of drug elimination half-life. *Philos. Trans. R. Soc. Lond. B. Biol. Sci*., **357,** 505–519.

Hastings, M.D., Porter, K.M., Maguire, J.D. *et al.* (2004). Dihydrofolate reductase mutations in *Plasmodium vivax* from Indonesia and therapeutic response to sulfadoxine plus pyrimethamine. *J. Infect. Dis*., **189,** 744–750.

Hay, S.I., Guerra, C.A., Tatem, A.J. *et al.* (2004). The global distribution and population at risk of malaria: past, present, and future. *Lancet Infect. Dis*., **4,** 327–336.

Hayton, K. and Su ,X.Z. (2004). Genetic and biochemical aspects of drug resistance in malaria parasites. *Curr. Drug Targets Infect. Disord*., **4,** 1–10.

Hellgren, U., Alvan, G., and Jerling, M. (1993). On the question of interindividual variations in chloroquine concentrations. *Eur. J. Clin. Pharmacol*., **45,** 383–385.

Hellgren, U., Ericsson, O., Kihamia, C.M. *et al.* (1994). Malaria parasites and chloroquine concentrations in Tanzanian schoolchildren. *Trop. Med. Parasitol*., **45,** 293–297.

Hellgren, U., Kihamia, C.M., Mahikwano, L.F. *et al.* (1989). Response of *Plasmodium falciparum* to chloroquine treatment: Relation to whole blood concentrations of chloroquine and desethylchloroquine. *Bull. World Health Organ*., **67,** 197–202.

Helsby, N.A., Edwards, G., Breckenridge, A.M. *et al.* (1993). The multiple dose pharmacokinetics of proguanil. *Br. J. Clin. Pharmacol*., **35,** 653–656.

Hoshen, M.B., Na-Bangchang, K., Stein, W.D., *et al.* (2000). Mathematical modeling of the chemotherapy of *Plasmodium falciparum* malaria with artesunate: Postulation of "dormancy," a partial cytostatic effect of the drug, and its implication for treatment regimens. *Parasitology*, **121,** 237–246.

Hoskins, J.M., Shenfield, G.M., and Gross, A.S. (1998). Relationship between proguanil metabolic ratio and CYP2C19 genotype in a Caucasian population. *Br. J. Clin. Pharmacol*., **46,** 499–504.

Hung, T.Y., Davis, T.M., Ilett, K.F. *et al.* (2004). Population pharmacokinetics of piperaquine in adults and children with uncomplicated falciparum or vivax malaria. Br. J. Clin. Pharmacol., **57,** 253–262.

Hyde, J.E. (1990). The dihydrofolate reductase–thymidylate synthetase gene in the drug resistance of malaria parasites. *Pharmacol. Ther*., **48,** 45–59.

Hyde, J.E. (2002). Mechanisms of resistance of *Plasmodium falciparum* to antimalarial drugs. *Microbes Infect*., **4,** 165–174.

Ilett, K.F., Ethell, B.T., Maggs, J.L. *et al.* (2002). Glucuronidation of dihydroartemisinin *in vivo* and by human liver microsomes and expressed UDP-glucuronosyltransferases. *Drug Metab. Dispos*., **30,** 1005–1012.

Imwong, M., Pukrittakayamee, S., Looareesuwan, S. *et al.* (2001). Association of genetic mutations in *Plasmodium vivax dhfr* with resistance to sulfadoxine-pyrimethamine: Geographical and clinical correlates. *Antimicrob. Agents Chemother*., **45,** 3122–3127.

Imwong, M., Pukrittayakamee, S., Renia, L. *et al.* (2003). Novel point mutations in the dihydrofolate reductase gene of *Plasmodium vivax*: Evidence for sequential selection by drug pressure. *Antimicrob. Agents Chemother*., **47,** 1514–1521.

Ingelman-Sundberg, M. (2001). Pharmacogenetics: an opportunity for a safer and more efficient pharmacotherapy. *J. Intern. Med*., **250,** 186–200.

Ingelman-Sundberg, M. (2004a). Human drug metabolising cytochrome P450 enzymes: Properties and polymorphisms. Naunyn Schmiedebergs. *Arch. Pharmacol*., **369,** 89–104.

Ingelman-Sundberg, M. (2004b). Pharmacogenetics of cytochrome P450 and its applications in drug therapy: The past, present and future. *Trends Pharmacol. Sci.*, **25,** 193–200.

Inselberg, J. (1985). Induction and isolation of artemisinin-resistant mutants of *Plasmodium falciparum*. *Am. J. Trop. Med. Hyg.*, **34,** 417–418.

Ittarat, W., Pickard, A.L., Rattanasinganchan, P. *et al.* (2003). Recrudescence in artesunate-treated patients with falciparum malaria is dependent on parasite burden not on parasite factors. *Am. J. Trop. Med. Hyg.*, **68,** 147–152.

Johnson, D.J., Fidock, D.A., Mungthin, M. *et al.* (2004). Evidence for a central role for PfCRT in conferring *Plasmodium falciparum* resistance to diverse antimalarial agents. *Mol, Cell,* **15,** 867–877.

Jung, M., Lee, K., Kim, H. *et al.* (2004). Recent advances in artemisinin and its derivatives as antimalarial and antitumor agents. *Curr. Med. Chem.*, **11,** 1265–1284.

Kain, K.C. (1995). Chemotherapy and prevention of drug-resistant malaria. *Wilderness Environ. Med.*, **6,** 307–324.

Kaneko, A., Bergqvist, Y., Takechi, M. *et al.* (1999). Intrinsic efficacy of proguanil against falciparum and vivax malaria independent of the metabolite cycloguanil. *J. Infect. Dis.*, **179,** 974–979.

Kaneko, A., Kaneko, O., Taleo, G. *et al.* (1997). High frequencies of CYP2C19 mutations and poor metabolism of proguanil in Vanuatu. *Lancet*, **349,** 921–922.

Karim, E.A., Ibrahim, K.E., Hassabalrasoul, M.A. *et al.* (1992). A study of chloroquine and desethylchloroquine plasma levels in patients infected with sensitive and resistant malaria parasites. *J. Pharm. Biomed. Anal.*, **10,** 219–223.

Karunajeewa, H., Lim, C., Hung, T.Y. *et al.* (2004). Safety evaluation of fixed combination piperaquine plus dihydroartemisinin (Artekin) in Cambodian children and adults with malaria. *Br. J. Clin. Pharmacol.*, **57,** 93–99.

Karunajeewa, H.A., Ilett, K.F., Dufall, K. *et al.* (2004). Disposition of artesunate and dihydroartemisinin after administration of artesunate suppositories in children from Papua New Guinea with uncomplicated malaria. *Antimicrob. Agents Chemother.*, **48,** 2966–2972.

Kawamoto, F., Liu, Q., Ferreira, M.U. *et al.* (1999). How prevalent are *Plasmodium ovale* and *P. malariae* in East Asia? *Parasitol. Today,* **15,** 422–426.

Ketrangsee, S., Vijaykadga, S., Yamokgul, P. *et al.* (1992). Comparative trial on the response of *Plasmodium falciparum* to halofantrine and mefloquine in Trat province, eastern Thailand. *Southeast Asian J. Trop. Med. Public. Health*, **23,** 55–58.

Kim, K.A., Park, J.Y., Lee, J.S. *et al.* (2003). Cytochrome P450 2C8 and CYP3A4/5 are involved in chloroquine metabolism in human liver microsomes. *Arch. Pharm. Res.*, **26,** 631–637.

Kofoed, P.E., Poulsen, A., Co, F. *et al.* (2003). No benefits from combining chloroquine with artesunate for 3 days for treatment of *Plasmodium falciparum* in Guinea-Bissau. *Trans. R. Soc. Trop. Med. Hyg.*, **97,** 429–433.

Koopmans, R., Duc, D.D., Kager, P.A. *et al.* (1998). The pharmacokinetics of artemisinin suppositories in Vietnamese patients with malaria. *Trans. R. Soc. Trop. Med. Hyg.*, **92,** 434–436.

Koopmans, R., Ha, L.D., Duc, D.D. *et al.* (1999). The pharmacokinetics of artemisinin after administration of two different suppositories to healthy Vietnamese subjects. *Am. J. Trop. Med. Hyg.*, **60,** 244–247.

Korsinczky, M., Chen, N., Kotecka, B. *et al.* (2000). Mutations in *Plasmodium falciparum* cytochrome b that are associated with atovaquone resistance are located at a putative drug-binding site. *Antimicrob. Agents Chemother.*, **44,** 2100–2108.

Korsinczky, M., Fischer, K., Chen, N. *et al.* (2004). Sulfadoxine resistance in *Plasmodium vivax* is associated with a specific amino acid in dihydropteroate synthase at the putative sulfadoxine-binding site. *Antimicrob. Agents Chemother.*, **48,** 2214–2222.

Krishna, S. and White, N.J. (1996). Pharmacokinetics of quinine, chloroquine and amodiaquine. Clinical implications. *Clin. Pharmacokinet.*, **30,** 263–299.

Krishna, S., Planche, T., Agbenyega, T. *et al.* (2001). Bioavailability and preliminary clinical efficacy of intrarectal artesunate in Ghanaian children with moderate malaria. *Antimicrob. Agents Chemother.*, **45,** 509–516.

Kumar, N. and Zheng, H. (1990). Stage-specific gametocytocidal effect *in vitro* of the antimalaria drug qinghaosu on *Plasmodium falciparum*. *Parasitol. Res.*, **76,** 214–218.

Lang, T. and Greenwood, B. (2003). The development of Lapdap, an affordable new treatment for malaria. *Lancet Infect. Dis.*, **3,** 162–168.

Laufer, M.K. and Plowe, C.V. (2004). Withdrawing antimalarial drugs: Impact on parasite resistance and implications for malaria treatment policies. *Drug Resist Updat.*, **7,** 279–288.

Le Bras, J. and Durand, R. (2003). The mechanisms of resistance to antimalarial drugs in *Plasmodium falciparum*. *Fundam. Clin. Pharmacol.*, **17,** 147–153.

Le Mire, J., Arnulf, L., and Guibert, P. (2004). Malaria: control strategies, chemoprophylaxis, diagnosis, and treatment. *Clin. Occup. Environ. Med.*, **4,** 143–165.

Le, T.A., Davis, T.M., Tran, Q.B. *et al.* (1997). Delayed parasite clearance in a splenectomized patient with falciparum malaria who was treated with artemisinin derivatives. *Clin. Infect. Dis.*, **25,** 923–925.

Leartsakulpanich, U., Imwong, M., Pukrittayakamee, S. *et al.* (2002). Molecular characterization of dihydrofolate reductase in relation to antifolate resistance in *Plasmodium vivax*. *Mol. Biochem. Parasitol.*, **119,** 63–73.

Lee, I.S. and Hufford, C.D. (1990). Metabolism of antimalarial sesquiterpene lactones. *Pharmacol. Ther.*, **48,** 345–355.

Lefevre, G., Bindschedler, M., Ezzet, F. *et al.* (2000). Pharmacokinetic interaction trial between coartemether and mefloquine. *Eur. J. Pharm. Sci.*, **10,** 141–151.

Li, X.Q., Bjorkman, A., Andersson, T.B. *et al.* (2002). Amodiaquine clearance and its metabolism to N-desethylamodiaquine is mediated by CYP2C8: A new high affinity and turnover enzyme-specific probe substrate. *J. Pharmacol. Exp. Ther.*, **300,** 399–407.

Li, X.Q., Bjorkman, A., Andersson, T.B. *et al.* (2003). Identification of human cytochrome P(450)s that metabolise antiparasitic drugs and predictions of *in vivo* drug hepatic clearance from *in vitro* data. *Eur. J. Clin. Pharmacol.*, **59,** 429–442.

Linder, M.W., Prough, R.A., and Valdes, R. Jr. (1997). Pharmacogenetics: a laboratory tool for optimizing therapeutic efficiency. *Clin. Chem.*, **43,** 254–266.

Little, J.M., Lester, R., Kuipers, F. *et al.* (1999). Variability of human hepatic UDP-glucuronosyltransferase activity. *Acta Biochim. Pol.*, **46,** 351–363.

Looareesuwan, S., Wilairatana, P., Krudsood, S. *et al.* (1999). Chloroquine sensitivity of *Plasmodium vivax* in Thailand. *Ann. Trop. Med. Parasitol.*, **93,** 225–230.

Lopes, D., Rungsihirunrat, K., Nogueira, F. *et al.* (2002). Molecular characterization of drug-resistant *Plasmodium falciparum* from Thailand. *Malar. J.*, **1,** 12.

Lowenthal, M.N. (1999). *Plasmodium ovale* in southern Africa. *Trans. R. Soc. Trop. Med. Hyg.*, **93,** 107.

Lu, A.H., Shu, Y., Huang, S.L. *et al.* (2000). *In vitro* proguanil activation to cycloguanil is mediated by CYP2C19 and CYP3A4 in adult Chinese liver microsomes. *Acta Pharmacol. Sin.*, **21,** 747–752.

Luxemburger, C., Brockman, A., Silamut, K. *et al.* (1998). Two patients with falciparum malaria and poor *in vivo* responses to artesunate. *Trans. R. Soc. Trop. Med. Hyg.*, **92,** 668–669.

Lysenko, A.J. and Beljaev, A.E. (1969). An analysis of the geographical distribution of *Plasmodium ovale*. *Bull. World Health Organ.*, **40,** 383–394.

Mackenzie, P.I., Gregory, P.A., Gardner-Stephen, D.A. *et al.* (2003). Regulation of UDP glucuronosyltransferase genes. *Curr. Drug Metab.*, **4,** 249–257.

Mackenzie, P.I., Miners, J.O., and McKinnon, R.A. (2000). Polymorphisms in UDP glucuronosyltransferase genes: functional consequences and clinical relevance. *Clin. Chem. Lab. Med.*, **38,** 889–892.

Mackenzie, P.I., Owens, I.S., Burchell, B. *et al.* (1997). The UDP glycosyltransferase gene superfamily: Recommended nomenclature update based on evolutionary divergence. *Pharmacogenetics*, **7,** 255–269.

Maguire, J.D., Sumawinata, I.W., Masbar, S. *et al.* (2002). Chloroquine-resistant *Plasmodium malariae* in south Sumatra, Indonesia. *Lancet*, **360,** 58–60.

Maitland, K., Bejon, P., and Newton, C.R. (2003). Malaria. *Curr. Opin. Infect. Dis.*, **16,** 389–395.

Maitland, K., Williams, T.N., Kotecka, B.M. *et al.* (1997). Plasma chloroquine concentrations in young and older malaria patients treated with chloroquine. *Acta Trop.*, **66,** 155–161.
Marlar-Than, Myat-Phone-Kyaw, Aye-Yu-Soe *et al.* (1995). Development of resistance to chloroquine by *Plasmodium vivax* in Myanmar. *Trans. R. Soc. Trop. Med. Hyg.*, **89,** 307–308.
Masimirembwa, C., Bertilsson, L., Johansson, I. *et al.* (1995). Phenotyping and genotyping of S mephenytoin hydroxylase (cytochrome P450 2C19) in a Shona population of Zimbabwe. *Clin. Pharmacol. Ther.*, **57,** 656–661.
Masimirembwa, C.M., Gustafsson, L.L., Dahl, M.L. *et al.* (1996). Lack of effect of chloroquine on the debrisoquine (CYP2D6) and S-mephenytoin (CYP2C19) hydroxylation phenotypes. *Br. J. Clin. Pharmacol.*, **41,** 344–346.
Masimirembwa, C.M., Thompson, R., and Andersson, T.B. (2001). *In vitro* high throughput screening of compounds for favorable metabolic properties in drug discovery. *Comb. Chem. High Throughput Screen.*, **4,** 245–263.
Masta, A., Lum, J.K., Tsukahara, T. *et al.* (2003). Analysis of Sepik populations of Papua New Guinea suggests an increase of CYP2C19 null allele frequencies during the colonization of Melanesia. *Pharmacogenetics*, **13,** 697–700.
May, D.G., Porter, J., Wilkinson, G.R. *et al.* (1994). Frequency distribution of dapsone N-hydroxylase, a putative probe for P4503A4 activity, in a white population. *Clin. Pharmacol. Ther.*, **55,** 492–500.
May, J. and Meyer, C.G. (2003). Association of *Plasmodium falciparum* chloroquine resistance transporter variant T76 with age-related plasma chloroquine levels. *Am. J. Trop. Med. Hyg.*, **68,** 143–146.
Mberu, E.K., Wansor, T., Sato, H. *et al.* (1995). Japanese poor metabolizers of proguanil do not have an increased risk of malaria chemoprophylaxis breakthrough. *Trans. R. Soc. Trop. Med. Hyg.*, **89,** 658–659.
McGready, R. and Nosten, F. (1999). The Thai–Burmese border: Drug studies of *Plasmodium falciparum* in pregnancy. *Ann. Trop. Med. Parasitol.*, **93,** S19–S23.
McKinnon, R.A. (2000). Cytochrome P450. 1. Multiplicity and function. *Aust. J. Hosp. Pharm.*, **30,** 54–56.
Meech, R. and Mackenzie, P.I. (1997). Structure and function of uridine diphosphate glucuronosyltransferases. *Clin. Exp. Pharmacol. Physiol.*, **24,** 907–915.
Meech, R. and Mackenzie, P.I. (1998). Determinants of UDP glucuronosyltransferase membrane association and residency in the endoplasmic reticulum. *Arch. Biochem. Biophys.*, **356,** 77–85.
Mehlotra, R.K., Lorry, K., Kastens, W. *et al.* (2000). Random distribution of mixed species malaria infections in Papua New Guinea. *Am. J. Trop. Med. Hyg.*, **62,** 225–231.
Mendis, K., Sina, B.J., Marchesini, P. *et al.* (2001). The neglected burden of *Plasmodium vivax* malaria. *Am. J. Trop. Med. Hyg.*, **64,** 97–106.
Mihara, K., Svensson, U.S., Tybring, G. *et al.* (1999). Stereospecific analysis of omeprazole supports artemisinin as a potent inducer of CYP2C19. *Fundam. Clin. Pharmacol.*, **13,** 671–675.
Miller, J.L. and Trepanier, L.A. (2002). Inhibition by atovaquone of CYP2C9-mediated sulphamethoxazole hydroxylamine formation. *Eur. J. Clin. Pharmacol.*, **58,** 69–72.
Miller, M.S., McCarver, D.G., Bell, D.A. *et al.* (1997). Genetic polymorphisms in human drug metabolic enzymes. *Fundam. Appl. Toxicol.*, **40,** 1–14.
Miners, J.O. and Birkett, D.J. (1998). Cytochrome P4502C9: An enzyme of major importance in human drug metabolism. *Br. J. Clin. Pharmacol.*, **45,** 525–538.
Miners, J.O. and Mackenzie, P.I. (1991). Drug glucuronidation in humans. *Pharmacol. Ther.*, **51,** 347–369.
Miners, J.O., McKinnon, R.A., and Mackenzie, P.I. (2002). Genetic polymorphisms of UDP-glucuronosyltransferases and their functional significance. *Toxicology*, **181–182,** 453–456.
Mita, T., Akira, K., Lum, J.K. *et al.* (2004). Expansion of wild-type allele rather than back mutation in *pfcrt* explains the recent recovery of chloroquine sensitivity of *Plasmodium falciparum* in Malawi. *Mol. Biochem. Parasitol.*, **135,** 159–163.
Mockenhaupt, F.P. (1995). Mefloquine resistance in *Plasmodium falciparum. Parasitol. Today*, **11,** 248–253.
Mockenhaupt, F.P., May, J., Bergqvist, Y. *et al.* (2000). Concentrations of chloroquine and malaria parasites in blood in Nigerian children. *Antimicrob. Agents Chemother.*, **44,** 835–839.

Mohapatra, P.K., Namchoom, N.S., Prakash, A. *et al.* (2003). Therapeutic efficacy of antimalarials in *Plasmodium falciparum* malaria in an Indo–Myanmar border area of Arunachal Pradesh. *Indian J. Med. Res.*, **118,** 71–76.

Monaghan, G., Clarke, D.J., Povey, S. *et al.* (1994). Isolation of a human YAC contig encompassing a cluster of *UGT2* genes and its regional localization to chromosome 4q13. *Genomics,* **23,** 496–499.

Mu, J., Ferdig, M.T., Feng, X. *et al.* (2003). Multiple transporters associated with malaria parasite responses to chloroquine and quinine. *Mol. Microbiol.*, **49,** 977–989.

Muehlen, M., Schreiber, J., Ehrhardt, S. *et al.* (2004). Short communication: Prevalence of mutations associated with resistance to atovaquone and to the antifolate effect of proguanil in *Plasmodium falciparum* isolates from northern Ghana. *Trop. Med. Int. Health,* **9,** 361–363.

Murphy, G.S., Basri, H., Purnomo *et al.* (1993). Vivax malaria resistant to treatment and prophylaxis with chloroquine. *Lancet,* **341,** 96–100.

Nagata, K. and Yamazoe, Y. (2002). Genetic polymorphism of human cytochrome P450 involved in drug metabolism. *Drug Metabol. Pharmacokin.*, **17,** 167–189.

Nagesha, H.S., Din-Syafruddin, Casey, G.J. *et al.* (2001). Mutations in the *pfmdr1*, *dhfr* and *dhps* genes of *Plasmodium falciparum* are associated with *in vivo* drug resistance in west Papua, Indonesia. *Trans. R. Soc. Trop. Med. Hyg.*, **95,** 43–49.

Nair, S., Williams, J.T., Brockman, A. *et al.* (2003). A selective sweep driven by pyrimethamine treatment in southeast asian malaria parasites. *Mol. Biol. Evol..* **20,** 1526–1536.

Navaratnam, V., Mansor, S.M., Sit, N.W. *et al.* (2000). Pharmacokinetics of artemisinin-type compounds. *Clin. Pharmacokinet.*, **39,** 255–270.

Nelson, D.R., Zeldin, D.C., Hoffman, S.M. *et al.* (2004). Comparison of cytochrome P450 (CYP) genes from the mouse and human genomes, including nomenclature recommendations for genes, pseudogenes and alternative-splice variants. *Pharmacogenetics*, **14,** 1–18.

Nielsen, T.L., Rasmussen, B.B., Flinois, J.P. *et al.* (1999). *In vitro* metabolism of quinidine: the (3S)-3-hydroxylation of quinidine is a specific marker reaction for cytochrome P-4503A4 activity in human liver microsomes. *J. Pharmacol. Exp. Ther.*, **289,** 31–37.

Nomura, T., Carlton, J.M., Baird, J.K. *et al.* (2001). Evidence for different mechanisms of chloroquine resistance in 2 *Plasmodium* species that cause human malaria. *J. Infect. Dis.*, **183,** 1653–1661.

Nosten, F. and Ashley, E. (2004). The detection and treatment of *Plasmodium falciparum* malaria: Time for change. *J. Postgrad. Med.*, **50,** 35–39.

Nosten, F., Hien, T.T., and White, N.J. (1998). Use of artemisinin derivatives for the control of malaria. *Med. Trop. (Mars)*, **58**, 45–49.

Nosten, F., McGready, R., Simpson, J.A. *et al.* (1999). Effects of *Plasmodium vivax* malaria in pregnancy. *Lancet*, **354,** 546–549.

Ochong, E.O., van den Broek, IV, Keus, K. *et al.* (2003). Short report: association between chloroquine and amodiaquine resistance and allelic variation in the *Plasmodium falciparum* multiple drug resistance 1 gene and the chloroquine resistance transporter gene in isolates from the upper Nile in southern Sudan. *Am. J. Trop. Med. Hyg.,* **69,** 184–187.

Oduola, A.M., Sowunmi, A., Milhous, W.K. *et al.* (1992). Innate resistance to new antimalarial drugs in *Plasmodium falciparum* from Nigeria. *Trans. R. Soc. Trop. Med. Hyg.*, **86,** 123–126.

Olliaro, P.L. and Taylor, W.R. (2004). Developing artemisinin based drug combinations for the treatment of drug resistant falciparum malaria: A review. *J. Postgrad. Med.*, **50,** 40–44.

Omari, A.A., Gamble, C., and Garner, P. (2004). Artemether-lumefantrine for uncomplicated malaria: A systematic review. *Trop. Med. Int. Health*, **9**, 192–199.

O'Neill, P.M., Scheinmann, F., Stachulski, A.V. *et al.* (2001). Efficient preparations of the beta-glucuronides of dihydroartemisinin and structural confirmation of the human glucuronide metabolite. *J. Med. Chem.*, **44,** 1467–1470.

Owens, I.S. and Ritter, J.K. (1995). Gene structure at the human *UGT1* locus creates diversity in isozyme structure, substrate specificity, and regulation. *Prog. Nucleic Acid Res. Mol. Biol.*, **51,** 305–338.

Payne, D. (1987). Spread of chloroquine resistance in *Plasmodium falciparum. Parasitol. Today*, **3,** 241–246.

Persson, I., Aklillu, E., Rodrigues, F. *et al.* (1996). S-mephenytoin hydroxylation phenotype and CYP2C19 genotype among Ethiopians. *Pharmacogenetics*, **6,** 521–526.

Peters, W. (1987). *Chemotherapy and Drug Resistance in Malaria*, 2nd ed. Academic, London, UK. p. 1091.

Peters, W. (1998). Drug resistance in malaria parasites of animals and man. *Adv. Parasitol.*, **41,** 1–62.

Peterson, D.S., Milhous, W.K., and Wellems, T.E. (1990). Molecular basis of differential resistance to cycloguanil and pyrimethamine in *Plasmodium falciparum* malaria. *Proc. Natl. Acad. Sci. USA*, **87,** 3018–3022.

Peterson, D.S., Walliker, D., and Wellems, T.E. (1988). Evidence that a point mutation in dihydrofolate reductase–thymidylate synthase confers resistance to pyrimethamine in falciparum malaria. *Proc. Natl. Acad. Sci. USA,* **85,** 9114–9118.

Pettinelli, F., Pettinelli, M.E., Eldin de Pecoulas, P. *et al.* (2004). Short report: High prevalence of multidrug-resistant *Plasmodium falciparum* malaria in the French territory of Mayotte. *Am. J. Trop. Med. Hyg.*, **70,** 635–637.

Phan, G.T., de Vries, P.J., Tran, B.Q. *et al.* (2002). Artemisinin or chloroquine for blood stage *Plasmodium vivax* malaria in Vietnam. *Trop. Med. Int. Health*, **7,** 858–864.

Phillips, E.J., Keystone, J.S., and Kain, K.C. (1996). Failure of combined chloroquine and high-dose primaquine therapy for *Plasmodium vivax* malaria acquired in Guyana, South America. *Clin. Infect. Dis.*, **23,** 1171–1173.

Pickard, A.L., Wongsrichanalai, C., Purfield, A. *et al.* (2003). Resistance to antimalarials in Southeast Asia and genetic polymorphisms in *pfmdr1*. *Antimicrob. Agents Chemother.*, **47,** 2418–2423.

Plowe, C.V., Cortese, J.F., Djimde, A. *et al.* (1997). Mutations in *Plasmodium falciparum* dihydrofolate reductase and dihydropteroate synthase and epidemiologic patterns of pyrimethamine–sulfadoxine use and resistance. *J. Infect. Dis.*, **176,** 1590–1596.

Ploypradith, P. (2004). Development of artemisinin and its structurally simplified trioxane derivatives as antimalarial drugs. *Acta Trop.*, **89,** 329–342.

Price, R.N. (2000). Artemisinin drugs: novel antimalarial agents. *Expert Opin. Investig. Drugs*, **9,** 1815–1827.

Projean, D., Baune, B., Farinotti, R. *et al.* (2003). *In vitro* metabolism of chloroquine: identification of CYP2C8, CYP3A4, and CYP2D6 as the main isoforms catalyzing N-desethylchloroquine formation. *Drug Metab. Dispos.*, **31,** 748–754.

Randrianarivelojosia, M., Raharimalala, L.A., Randrianasolo, L. *et al.* (2001). Madagascan isolates of *Plasmodium falciparum* showing low sensitivity to artemether *in vitro. Ann. Trop. Med. Parasitol.*, **95,** 237–243.

Reed, M.B., Saliba, K.J., Caruana, S.R. *et al.* (2000). Pgh1 modulates sensitivity and resistance to multiple antimalarials in *Plasmodium falciparum. Nature*, **403,** 906–909.

Ringwald, P., Bickii, J., and Basco, L.K. (1999). *In vitro* activity of dihydroartemisinin against clinical isolates of *Plasmodium falciparum* in Yaounde, Cameroon. *Am. J. Trop. Med. Hyg.*, **61,** 187–192.

Ringwald, P., Same Ekobo, A., Keundjian, A. *et al.* (2000). Chemoresistance of *P. falciparum* in urban areas of Yaounde, Cameroon. Part 1: Surveillance of *in vitro* and *in vivo* resistance of *Plasmodium falciparum* to chloroquine from 1994 to 1999 in Yaounde, Cameroon. *Trop. Med. Int. Health*, **5,** 612–619.

Ritter, J.K., Chen, F., Sheen, Y.Y. *et al.* (1992). A novel complex locus *UGT1* encodes human bilirubin, phenol, and other UDP-glucuronosyltransferase isozymes with identical carboxyl termini. *J. Biol. Chem.*, **267,** 3257–3261.

Rodrigues, A.D. (1999). Integrated cytochrome P450 reaction phenotyping: attempting to bridge the gap between cDNA-expressed cytochromes P450 and native human liver microsomes. *Biochem. Pharmacol.*, **57,** 465–480.

Schwartz, E., Bujanover, S., and Kain, K.C. (2003). Genetic confirmation of atovaquone-proguanil-resistant *Plasmodium falciparum* malaria acquired by a nonimmune traveler to East Africa. *Clin. Infect. Dis.*, **37,** 450–451.

Schwartz, E., Regev-Yochay, G., and Kurnik, D. (2000). Short report: A consideration of primaquine dose adjustment for radical cure of *Plasmodium vivax* malaria. *Am. J. Trop. Med. Hyg.*, **62,** 393–395.

Scopel, K.K., Fontes, C.J., Nunes, A.C. *et al.* (2004). High prevalence of *Plamodium malariae* infections in a Brazilian Amazon endemic area (Apiacas–Mato Grosso State) as detected by polymerase chain reaction. *Acta Trop.*, **90**, 61–64.

Sen, S. and Ferdig, M. (2004). QTL analysis for discovery of genes involved in drug responses. *Curr. Drug Targets Infect. Disord.*, **4**, 53–63.

Shmuklarsky, M.J., Klayman, D.L., Milhous, W.K. *et al.* (1993). Comparison of beta-artemether and beta-arteether against malaria parasites *in vitro* and *in vivo. Am. J. Trop. Med. Hyg.*, **48**, 377–384.

Sibley, C.H., Hyde, J.E., Sims, P.F. *et al.* (2001). Pyrimethamine–sulfadoxine resistance in *Plasmodium falciparum*: What next? *Trends Parasitol.*, **17**, 582–588.

Sidhu, A.B., Verdier-Pinard, D., and Fidock, D.A. (2002). Chloroquine resistance in *Plasmodium falciparum* malaria parasites conferred by *pfcrt* mutations. *Science*, **298**, 210–213.

Simonsson, U.S., Jansson, B., Hai, T.N. *et al.* (2003). Artemisinin autoinduction is caused by involvement of cytochrome P450 2B6 but not 2C9. *Clin. Pharmacol. Ther.*, **74**, 32–43.

Simooya, O.O., Sijumbil, G., Lennard, M.S. *et al.* (1998). Halofantrine and chloroquine inhibit CYP2D6 activity in healthy Zambians. *Br. J. Clin. Pharmacol.*, **45**, 315–317.

Sims, P., Wang, P., and Hyde, J.E. (1999). Selection and synergy in *Plasmodium falciparum. Parasitol. Today*, **15**, 132–134.

Singh, B., Kim Sung, L., Matusop. A. *et al.* (2004). A large focus of naturally acquired *Plasmodium knowlesi* infections in human beings. *Lancet*, **363**, 1017–1024.

Sirima, S.B., Tiono, A.B., Konate, A. *et al.* (2003). Efficacy of artesunate plus chloroquine for the treatment of uncomplicated malaria in children in Burkina Faso: A double-blind, randomized, controlled trial. *Trans. R. Soc. Trop. Med. Hyg.,* **97**, 345–349.

Skjelbo, E., Mutabingwa, T.K., Bygbjerg, I. *et al.* (1996). Chloroguanide metabolism in relation to the efficacy in malaria prophylaxis and the S-mephenytoin oxidation in Tanzanians. *Clin. Pharmacol. Ther.*, **59**, 304–311.

Smoak, B.L., DeFraites, R.F., Magill, A.J. *et al.* (1997). *Plasmodium vivax* infections in U.S. Army troops: Failure of primaquine to prevent relapse in studies from Somalia. *Am. J. Trop. Med. Hyg.,* **56**, 231–234.

Snewin, V.A., England, S.M., Sims, P.F. *et al.* (1989). Characterisation of the dihydrofolate reductase–thymidylate synthetase gene from human malaria parasites highly resistant to pyrimethamine. *Gene*, **76**, 41–52.

Snow, R.W., Trape, J.F., and Marsh, K. (2001). The past, present and future of childhood malaria mortality in Africa. *Trends Parasitol.*, **17**, 593–597.

Somogyi, A.A., Reinhard, H.A., and Bochner, F. (1996). Pharmacokinetic evaluation of proguanil: A probe phenotyping drug for the mephenytoin hydroxylase polymorphism. *Br. J. Clin. Pharmacol.*, **41**, 175–179.

Soto, J., Toledo, J., Gutierrez, P. *et al.* (2001). *Plasmodium vivax* clinically resistant to chloroquine in Colombia. *Am. J. Trop. Med. Hyg.*, **65**, 90–93.

Srivastava, I.K. and Vaidya, A.B. (1999). A mechanism for the synergistic antimalarial action of atovaquone and proguanil. *Antimicrob. Agents Chemother*, **43**, 1334–1339.

Steketee, R.W., Nahlen, B.L., Parise, M.E. *et al.* (2001). The burden of malaria in pregnancy in malaria-endemic areas. *Am. J. Trop. Med. Hyg.*, **64**, 28–35.

Stormer, E., von Moltke, L.L., and Greenblatt, D.J. (2000). Scaling drug biotransformation data from cDNA-expressed cytochrome P-450 to human liver: A comparison of relative activity factors and human liver abundance in studies of mirtazapine metabolism. *J. Pharmacol. Exp. Ther.*, **295**, 793–801.

Suh, K.N., Kain, K.C., and Keystone, J.S. (2004). Malaria. *CMAJ.* **170**, 1693–1702.

Svensson, U.S. and Ashton, M. (1999). Identification of the human cytochrome P450 enzymes involved in the *in vitro* metabolism of artemisinin. *Br. J. Clin. Pharmacol.*, **48**, 528–535.

Svensson, U.S., Ashton, M., Trinh, N.H. *et al.* (1998). Artemisinin induces omeprazole metabolism in human beings. *Clin. Pharmacol. Ther.*, **64**, 160–167.

Tahar, R., Ringwald, P., and Basco, L.K. (1998). Heterogeneity in the circumsporozoite protein gene of *Plasmodium malariae* isolates from subSaharan Africa. *Mol. Biochem. Parasitol.*, **92**, 71–78.

Talisuna, A.O., Bloland, P., and D'Alessandro, U. (2004). History, dynamics, and public health importance of malaria parasite resistance. *Clin. Microbiol. Rev.*, **17,** 235–254.

Taylor, W.R. and White, N.J. (2004). Antimalarial drug toxicity: A review. *Drug Saf.*, **27,** 25–61.

ter Kuile, F., White, N.J., Holloway, P. *et al.* (1993). *Plasmodium falciparum*: *In vitro* studies of the pharmacodynamic properties of drugs used for the treatment of severe malaria. *Exp. Parasitol.*, **76,** 85–95.

ter Kuile, F.O., Dolan, G., Nosten, F. *et al.* (1993). Halofantrine versus mefloquine in treatment of multidrug-resistant falciparum malaria. *Lancet*, **341,** 1044–1049.

Thapar, M.M., Ashton, M., Lindegardh, N. *et al.* (2002). Time-dependent pharmacokinetics and drug metabolism of atovaquone plus proguanil (Malarone) when taken as chemoprophylaxis. *Eur. J. Clin. Pharmacol.*, **58,** 19–27.

Tobian, A.A., Mehlotra, R.K., Malhotra, I. *et al.* (2000). Frequent umbilical cord-blood and maternal-blood infections with *Plasmodium falciparum, P. malariae,* and *P. ovale* in Kenya. *J. Infect. Dis.*, **182,** 558–563.

Tran, T.H., Dolecek, C., Pham, P.M. *et al.* (2004). Dihydroartemisinin-piperaquine against multidrug-resistant *Plasmodium falciparum* malaria in Vietnam: randomized clinical trial. *Lancet*, **363,** 18–22.

Tredger, J.M. and Stoll, S. (2002). Cytochrome P450 their impact on drug treatment. *Hospital Pharmacy*, **9,** 167–173.

Treeprasertsuk, S., Viriyavejakul, P., Silachamroon, U. *et al.* (2000). Is there any artemisinin resistance in falciparum malaria? *Southeast Asian J. Trop. Med. Public Health*, **31,** 825–828.

Triglia ,T. and Cowman, A.F. (1994). Primary structure and expression of the dihydropteroate synthetase gene of *Plasmodium falciparum. Proc. Natl. Acad. Sci. USA.*, **91,** 7149–7153.

Triglia, T., Menting, J.G., Wilson, C. *et al.* (1997). Mutations in dihydropteroate synthase are responsible for sulfone and sulfonamide resistance in *Plasmodium falciparum. Proc. Natl. Acad. Sci. USA*, **94,** 13,944–13,949.

Triglia, T., Wang, P., Sims, P.F. *et al.* (1998). Allelic exchange at the endogenous genomic locus in *Plasmodium falciparum* proves the role of dihydropteroate synthase in sulfadoxine-resistant malaria. *EMBO. J.*, **17,** 3807–3815.

Turgeon, D., Carrier, J.S., Levesque, E. *et al.* (2000). Isolation and characterization of the human *UGT2B15* gene, localized within a cluster of *UGT2B* genes and pseudogenes on chromosome 4. *J. Mol. Biol.*, **295,** 489–504.

Ukpe, I.S. (1998). *Plasmodium ovale* in South Africa. *Trans. R. Soc. Trop. Med. Hyg.*, **92,** 574.

van Agtmael, M.A., Gupta, V., van der Graaf, C.A., and van Boxtel, C.J. (1999a). The effect of grapefruit juice on the time-dependent decline of artemether plasma levels in healthy subjects. *Clin. Pharmacol. Ther.*, **66,** 408–414.

van Agtmael, M.A., Gupta, V., van der Wosten, T.H. *et al.* (1999b). Grapefruit juice increases the bioavailability of artemether. *Eur. J. Clin. Pharmacol.*, **55,** 405–410.

van Agtmael, M.A., Van Der Graaf, C.A., Dien, T.K. *et al.* (1998). The contribution of the enzymes CYP2D6 and CYP2C19 in the demethylation of artemether in healthy subjects. *Eur. J. Drug Metab. Pharmacokinet.*, **23,** 429–436.

Vieira, P.P., Ferreira, M.U., Das Gracas Alecrim, M. *et al.* (2004). *pfcrt* Polymorphism and the Spread of Chloroquine Resistance in *Plasmodium falciparum* Populations across the Amazon Basin. *J. Infect. Dis.*, **190,** 417–424.

Volkman, S. and Wirth, D. (1998). Functional analysis of *pfmdr1* gene of *Plasmodium falciparum. Methods Enzymol.*, **292,** 174–181.

Walker, O., Dawodu, A.H., Adeyokunnu, A.A. *et al.* (1983). Plasma chloroquine and desethylchloroquine concentrations in children during and after chloroquine treatment for malaria. *Br. J. Clin. Pharmacol.*, **16,** 701–705.

Wang, P., Lee, C.S., Bayoumi, R. *et al.* (1997). Resistance to antifolates in *Plasmodium falciparum* monitored by sequence analysis of dihydropteroate synthetase and dihydrofolate reductase alleles in a large number of field samples of diverse origins. *Mol. Biochem. Parasitol.*, **89,** 161–177.

Wang, P., Read, M., Sims, P.F. *et al.* (1997). Sulfadoxine resistance in the human malaria parasite *Plasmodium falciparum* is determined by mutations in dihydropteroate synthetase and an additional factor associated with folate utilization. *Mol. Microbiol.*, **23,** 979–986.

Wang, P., Wang, Q., Aspinall, T.V. *et al.* (2004). Transfection studies to explore essential folate metabolism and antifolate drug synergy in the human malaria parasite *Plasmodium falciparum*. *Mol. Microbiol.*, **51,** 1425–1438.

Wanwimolruk, S., Bhawan, S., Coville, P.F. *et al.* (1998). Genetic polymorphism of debrisoquine (CYP2D6) and proguanil (CYP2C19) in South Pacific Polynesian populations. *Eur. J. Clin. Pharmacol.*, **54,** 431–435.

Wanwimolruk, S., Pratt, E.L., Denton, J.R. *et al.* (1995). Evidence for the polymorphic oxidation of debrisoquine and proguanil in a New Zealand Maori population. *Pharmacogenetics*, **5,** 193–198.

Wanwimolruk, S., Wong, S.M., Zhang, H. *et al.* (1995). Metabolism of quinine in man: identification of a major metabolite, and effects of smoking and rifampicin pretreatment. *J. Pharm. Pharmacol.*, **47,** 957–963.

Ward, S.A., Helsby, N.A., Skjelbo, E. *et al.* (1991). The activation of the biguanide antimalarial proguanil cosegregates with the mephenytoin oxidation polymorphism—a panel study. *Br. J. Clin. Pharmacol.*, **31,** 689–692.

Watkins, W.M., Mberu, E.K., Nevill, C.G. *et al.* (1990). Variability in the metabolism of proguanil to the active metabolite cycloguanil in healthy Kenyan adults. *Trans. R. Soc. Trop. Med. Hyg.*, **84,** 492–495.

Weinshilboum, R. (2003). Inheritance and drug response. *N. Engl. J. Med.*, **348,** 529–537.

Wellems, T.E. and Plowe, C.V. (2001). Chloroquine-resistant malaria. *J. Infect. Dis.*, **184,** 770–776.

Wellems, T.E., Walker-Jonah, A., and Panton, L.J. (1991). Genetic mapping of the chloroquine-resistance locus on *Plasmodium falciparum* chromosome 7. *Proc. Natl. Acad. Sci. US.*, **88,** 3382–3386.

Wells, P.G., Mackenzie, P.I., Chowdhury, J.R. *et al.* (2004). Glucuronidation and the UDP-glucuronosyltransferases in health and disease. *Drug. Metab. Dispos.*, **32,** 281–290.

Wen, X., Wang, J.S., Backman, J.T. *et al.* (2002). Trimethoprim and sulfamethoxazole are selective inhibitors of CYP2C8 and CYP2C9, respectively. *Drug Metab. Dispos.*, **30,** 631–635.

Wernsdorfer, W.H. (1992). The biological and epidemiological basis of drug resistance in malaria parasites. *Southeast Asian J. Trop. Med. Public Health*, **23,** 123–129.

Wernsdorfer, W.H. and Noedl, H. (2003). Molecular markers for drug resistance in malaria: Use in treatment, diagnosis and epidemiology. *Curr. Opin. Infect. Dis.*, **16,** 553–558.

Wernsdorfer, W.H. (1994). Epidemiology of drug resistance in malaria. *Acta Trop.*, **56,** 143–156.

Wernsdorfer, W.H. and Payne, D. (1991). The dynamics of drug resistance in *Plasmodium falciparum*. *Pharmacol. Ther.*, **50,** 95–121.

White, N.J. (1994). Clinical pharmacokinetics and pharmacodynamics of artemisinin and derivatives. *Trans. R. Soc. Trop. Med. Hyg.*, **88,** S41–43.

White, N.J. (1999). Delaying antimalarial drug resistance with combination chemotherapy. *Parassitologia*, **41,** 301–308.

White, N.J. (2003). Malaria. In: G.C. Cook and A. Zumla (eds), *Manson's Tropical Diseases,* 21st ed. W.B. Saunders. pp. 1205–1295.[j15]

White, N.J. (2004). Antimalarial drug resistance. *J. Clin. Invest.*, **113,** 1084–1092.

White, N.J. and Olliaro, P. (1998). Artemisinin and derivatives in the treatment of uncomplicated malaria. *Med. Trop. (Mars)*, **58,** 54–56.

Wichmann, O., Betschart, B., Loscher, T. *et al.* (2003). Prophylaxis failure due to probable mefloquine resistant *P. falciparum* from Tanzania. *Acta. Trop.*, **86,** 63–65.

Wichmann, O., Jelinek, T., Peyerl-Hoffmann, G. *et al.* (2003). Molecular surveillance of the antifolate-resistant mutation I164L in imported African isolates of *Plasmodium falciparum* in Europe: Sentinel data from TropNetEurop. *Malar. J.*, **2,** 17.

Wichmann, O., Muehlen, M., Gruss, H. *et al.* (2004). Malarone treatment failure not associated with previously described mutations in the cytochrome *b* gene. *Malar. J.* **3,** 14.

Wilairatana, P., Silachamroon, U., Krudsood, S. *et al.* (1999). Efficacy of primaquine regimens for primaquine-resistant *Plasmodium vivax* malaria in Thailand. *Am. J. Trop. Med. Hyg.*, **61,** 973–977.

Wilson, C.M., Volkman, S.K., Thaithong, S. *et al.* (1993). Amplification of *pfmdr1* associated with mefloquine and halofantrine resistance in *Plasmodium falciparum* from Thailand. *Mol. Biochem. Parasitol.*, **57,** 151–160.

Win, T.T., Lin, K., Mizuno, S. *et al.* (2002). Wide distribution of *Plasmodium ovale* in Myanmar. *Trop. Med. Int. Health*, **7,** 231–239.

Winstanley, P (2001). Chlorproguanil-dapsone (LAPDAP) for uncomplicated falciparum malaria. *Trop. Med. Int. Health*, **6,** 952–954.

Winstanley, P.A. (2000). Chemotherapy for falciparum malaria: The armoury, the problems, and the prospects. *Parasitol. Today,* **16,** 146–153.

Winstanley, P.A. and Watkins, W.M. (1992). Pharmacology and parasitology: Integrating experimental methods and approaches to falciparum malaria. *Br. J. Clin. Pharmacol.*, **33,** 575–581.

Winstanley, P.A., Ward, S.A., and Snow, R.W. (2002). Clinical status and implications of antimalarial drug resistance. *Microbes Infect.*, **4,** 157–164.

Winter, H.R., Wang, Y., and Unadkat, J.D. (2000). CYP2C8/9 mediate dapsone N-hydroxylation at clinical concentrations of dapsone. *Drug Metab. Dispos.*, **28,** 865–868.

Wongsrichanalai, C., Nguyen, T.D., Trieu, N.T. *et al.* (1997). *In vitro* susceptibility of *Plasmodium falciparum* isolates in Vietnam to artemisinin derivatives and other antimalarials. *Acta Trop.*, **63,** 151–158.

Wongsrichanalai, C., Pickard, A.L., Wernsdorfer, W.H. *et al.* (2002). Epidemiology of drug-resistant malaria. *Lancet Infect. Dis.*, **2,** 209–218.

Wrighton, S.A., Brian, W.R., Sari, M.A. *et al.* (1990). Studies on the expression and metabolic capabilities of human liver cytochrome P450IIIA5 (HLp3). *Mol. Pharmacol.*, **38,** 207–213.

Wu, Y., Kirkman, L.A., and Wellems, T.E. (1996). Transformation of *Plasmodium falciparum* malaria parasites by homologous integration of plasmids that confer resistance to pyrimethamine. *Proc. Natl. Acad. Sci. USA.*, **93,** 1130–1134.

Yang, H., Liu, D., Yang, Y. *et al.* (2003). Changes in susceptibility of *Plasmodium falciparum* to artesunate *in vitro* in Yunnan Province, China. *Trans. R. Soc. Trop. Med. Hyg.*, **97,** 226–228.

Yusuf, I., Djojosubroto, M.W., Ikawati, R. *et al.* (2003). Ethnic and geographical distributions of CYP2C19 alleles in the populations of Southeast Asia. *Adv. Exp. Med. Biol.*, **531,** 37–46.

Yuthavong, Y. (2002). Basis for antifolate action and resistance in malaria. *Microbes Infect.*, **4,** 175–182.

Zhang, H., Coville, P.F., Walker, R.J. *et al.* (1997). Evidence for involvement of human CYP3A in the 3-hydroxylation of quinine. *Br. J. Clin. Pharmacol.*, **43,** 245–252.

Zhang, S.Q., Hai, T.N., Ilett, K.F. *et al.* (2001). Multiple dose study of interactions between artesunate and artemisinin in healthy volunteers. *Br. J. Clin. Pharmacol.*, **52,** 377–385.

Zhao, X.J., Yokoyama, H., Chiba, K. *et al.* (1996). Identification of human cytochrome P450 isoforms involved in the 3-hydroxylation of quinine by human live microsomes and nine recombinant human cytochromes P450. *J. Pharmacol. Exp. Ther.*, **279,** 1327–1334.

Zolg, J.W., Plitt, J.R., Chen, G.X. *et al.* (1989). Point mutations in the dihydrofolate reductase–thymidylate synthase gene as the molecular basis for pyrimethamine resistance in *Plasmodium falciparum. Mol. Biochem. Parasitol.*, **36,** 253–262.

Zuidema, J., Hilbers-Modderman, E.S., and Merkus, F.W. (1986). Clinical pharmacokinetics of dapsone. *Clin. Pharmacokinet.*, **11,** 299–315.

Evolutionary Origins of Human Malaria Parasites

Stephen M. Rich[a] **and Francisco J. Ayala**[b]

1. The Phylum Apicomplexa

The genus *Plasmodium* consists of nearly 200 described species that are parasitic to reptiles, birds, and mammals. *Plasmodium* belongs to the phylum Apicomplexa, which includes more than 5000 described species and many more still to be described. The Apicomplexa are all parasites, characterized by their eponym structure, the apical complex, which probably plays an important role in the parasite's penetration of host cells. Apicomplexa other than *Plasmodium* that cause disease in humans include *Toxoplasma*, *Cryptosporidium*, and *Babesia*, although the toll attributed to any one of these is small compared to *Plasmodium*. There is no fossil record of apicomplexans (Margulis *et al.*, 1993); molecular phylogenetic investigations provide the best insight to the origins of the phylum. These studies have indicated that the Apicomplexa are very ancient, perhaps as old as the multicellular kingdoms of plants, fungi, and animals, and thus somewhat older than one billion years (Ayala *et al.*, 1998). The *Plasmodium* lineage diverged from other Apicomplexa several hundred million years ago, perhaps earlier than the Cambrian and before the vertebrates (chordates) originated form their ancestral invertebrate lineage (Figure 6.1).

Some Apicomplexans are *monogenetic,* parasitic to one or several related species, without requiring an intermediate host; e.g., the coccidean *Cryptosporidium parvum*, which infects animal gut epitelia; although other coccideans, such as *Sarcocystis* and *Toxoplasma* are *digenetic* parasites, requiring two separate host species to complete their life cycle. The

---

Stephen M. Rich • Department of Plant, Soil and Insect Sciences, University of Massachusetts, Amherst, MA 01002, USA. **Francisco J. Ayala** • Department of Ecology and Evolutionary Biology, University of California, Irvine, California 92697, USA

Malaria: Genetic and Evolutionary Aspects, edited by Krishna R. Dronamraju and Paolo Arese, Springer, New York, 2006.

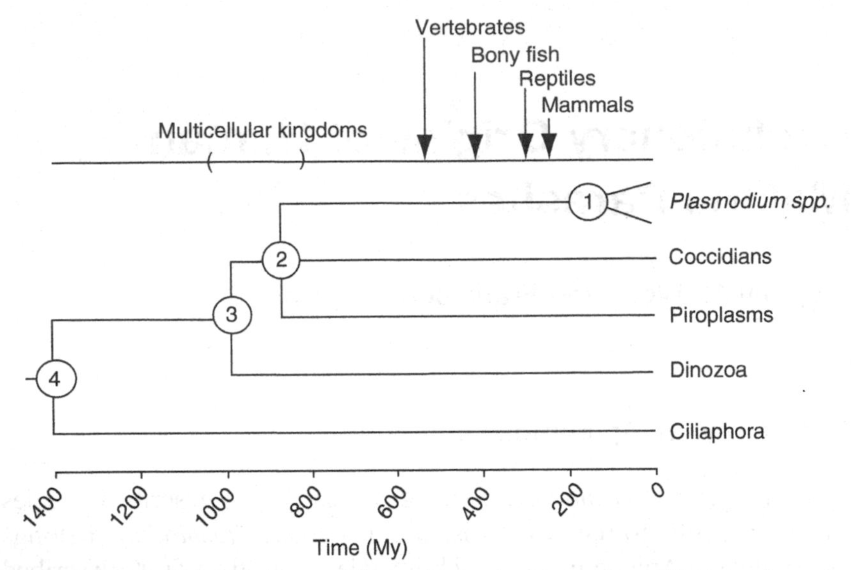

Figure 6.1. Simplified phylogeny of *Plasmodium* and related protozoa. The branching nodes refer to: 1, radiation of the *Plasmodium* genus; 2, radiation of the phylum Apicomplexa; 3 and 4, Apicomplexa divergence from two related phyla, Dinozoa and Ciliophora.

two host species of digenetic parasites are distinguished as being either definitive or intermediate, with the definitive host being the site in which sexual reproduction of the parasite occurs. All *Plasmodium* species are digenetic. An unsettled question is whether *Plasmodium* parasites evolved directly from monogenetic parasites of ancient marine invertebrates ancestors of modern chordates, or whether they descended from other digenetic parasites (Huff, 1938; Manwell, 1955; Garnham, 1966; Barta, 1989). Molecular phylogenetic comparisons have not resolved this issue, but it is apparent from analyses of ribosomal DNA and other genetic loci that the digenetic life style has multiple independent origins among apicomplexans (Barta, 1989; Escalante and Ayala, 1995; Fast *et al.*, 2002).

Vertebrates are the typical intermediate hosts of *Plasmodium*, while invertebrate species are the definitive hosts or vectors. *Plasmodium* are intracellular parasites occupying the blood cells of the intermediate host for a large part of their life cycle; accordingly, their invertebrate vectors are blood-feeding organisms, most typically mosquitoes, although in the case of some of the reptilian malarias, sand flies serve as the vector (Kimsey, 1992).

Two phyla related to the Apicomplexa are the Ciliophora (ciliates), which includes *Paramecium,* and the Dinozoa, which includes the dinoflagellates. Two large classes of Apicomplexa are the Coccidea, already mentioned, and the Hematozoa, which includes the order

Haemosporida, to which *Plasmodium* belongs, and the Piroplasmida, which includes species parasitic to dogs, cats, horses, and cattle and are mostly transmitted by ticks (definitive hosts).

2. The Genus Plasmodium

Human malaria is caused by four *Plasmodium* species: *P. falciparum, P. vivax, P. malariae,* and *P. ovale.* The human toll of malaria is stunning, perhaps the greatest of all human afflictions (Sherman, 1998), although nowadays AIDS and tuberculosis may approach the number of fatalities attributed to malaria, between 1 and 3 million deaths per year, mostly children. The worldwide incidence of malaria is estimated at 300–500 million cases per year. Most morbidity and mortality occurs in subSaharan Africa, caused by *P. falciparum*, the most virulent species, accounting for nearly 85% of the total. *P. vivax* is the most geographically widespread of the human malarias, estimated to account for 70–80 million clinical cases across much of Asia, Central and South America, the Middle East, and parts of Africa.

Various molecular phylogenetic analyses have revealed the relationships among the *Plasmodium* species. Herein, we show a representative phylogenetic tree based on the circumsporozoite protein (see, Figure 6.2) gene sequences. Table 6.1 provides information about the hosts and the geographic distribution of the species. The trees are obtained by the "neighbor-joining" (NJ) method (Saitou and Nei, 1987) based on genetic distances calculated according to Tamura's three-parameter method (Tamura, 1992). Trees obtained with other methods (such as maximum likelihood) and/or based on other measures of genetic distance have fundamentally identical topologies as those shown in Figure 6.2 (certainly with respect to the conclusions that will be formulated later). The root of the *Csp* tree has been determined by maximum likelihood. (Additional details can be found in Escalante and Ayala, 1995; Escalante *et al.*, 1995; and Ayala *et al.*, 1998). Estimates of divergence times based on two genes, *Csp* and *rRNA* trees are given in Table 6.2.

The phylogenies represented in Figure 6.2 include three human parasites, *P. falciparum, P. vivax,* and *P. malariae*. All four species (i.e., including *P. ovale*) have been included in phylogenies based on the mitochondrial gene encoding cytochrome-b (*cyt-B*) (Figure 6.3, after Perkins and Schall, 2002; Qari *et al.* 1996). These phylogenetic studies yield the following conclusions concerning the evolutionary history of the human malarial parasites.

1. The four human parasites, *P. falciparum, P. ovale, P. malariae,* and *P. vivax* are very remotely related to each other, so that the

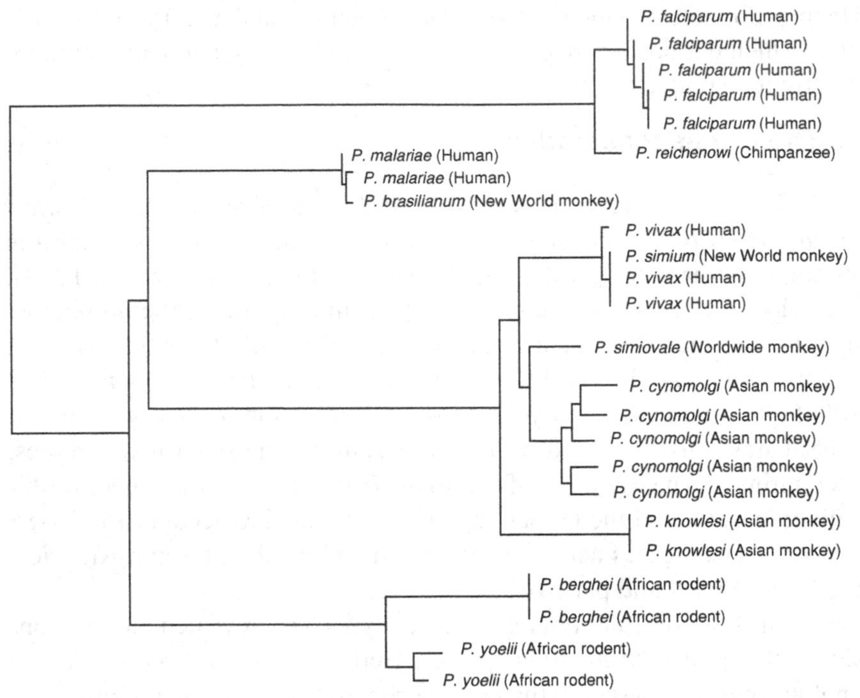

Figure 6.2. Phylogeny of 11 *Plasmodium* species (30 isolates) inferred from *Csp* gene sequences. Each parasite's host is given in parentheses. Some independent isolates have identical sequences of which only one is shown; thus, only five out of eight isolates are shown for *P. falciparum*, three out of four for *P. vivax*, and one out of two for *P. simium*.

evolutionary divergence of these four human parasites greatly predates the origin of the hominids. It follows that their parasitic associations with humans are phylogenetically independent (i.e., all but one – at the most) of these species have been laterally

Table 6.1. *Plasmodium* Species Used for Construction of Phylogeny Based on *Csp* Sequences

Species	Number of strains	Host	Geographic distribution
P. falciparum	8	Human	Tropics worldwide
P. malariae	2	Human	Tropics worldwide
P. vivax	4	Human	Tropics worldwide
P. reichenowi	1	Chimpanzee	African tropics
P. brasilianum	1	Monkey	New World tropics
P. simiovale	1	Monkey	Tropics worldwide
P. cynomolgi	5	Monkey	Asian tropics
P. simium	2	Monkey	New World tropics
P. knowlesi	2	Monkey	Asian tropics
P. berghei	2	Rodent	African tropics
P. yoelii	2	Rodent	Africa

Table 6.2. Time (in Million Years) of Divergence Between *Plasmodium* Species, Based on Genetic Distances at Two Loci (Ayala *et. al.*, 1999)

	rRNA	*Csp*
falciparum versus *reichenowi*	11.2 ± 2.5	8.9 ± 0.4
vivax versus monkey*	20.9 ± 3.8	25.2 ± 2.1
vivax versus *malariae*	75.7 ± 8.8	103.5 ± 0.6
falciparum versus *vivax/malariae*	75.7 ± 8.8	165.4 ± 1.6

**brasilianum* and *simium* not included

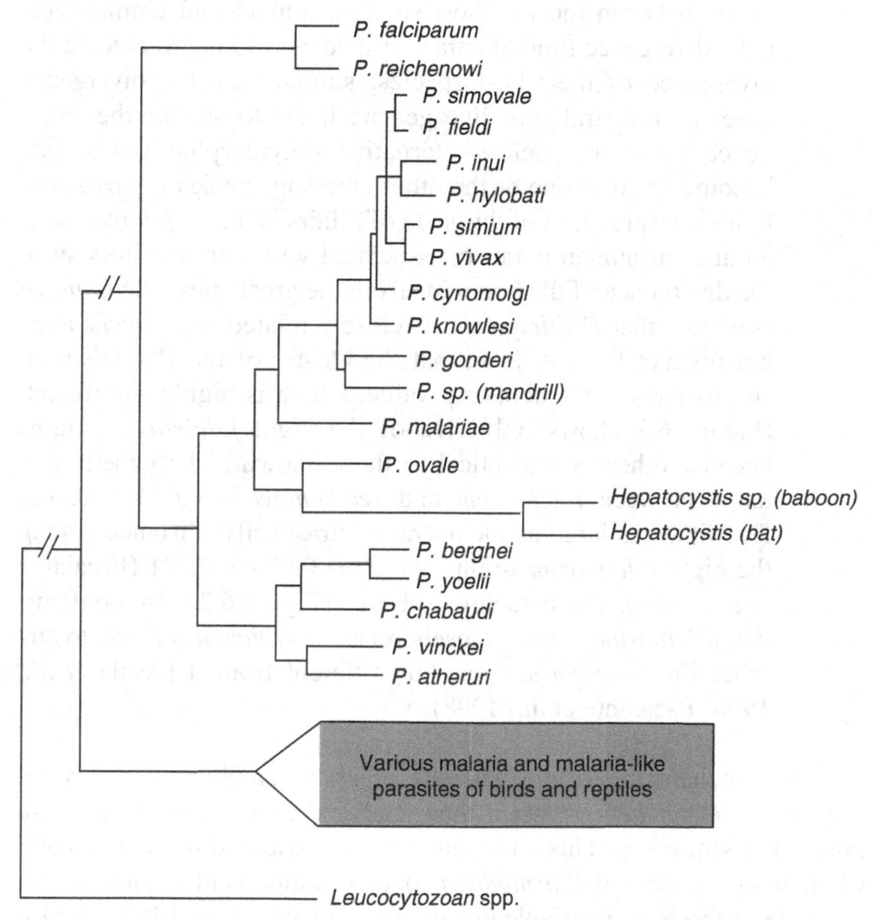

Figure 6.3. Phylogeny of 19 *Plasmodium* species (+2 *Hepatocystis* spp.) inferred from mitochondrial cytochrome-b gene sequences. "*Plasmodium sp.*" is an undescribed species isolated from a mandrill, *Mandrillus leucophaeus*, in Gabon. Maximum likelihood analyses give strong support for monophyly (i.e., a single common ancestor) of the mammalian malaria species. The inclusion of two *Hepatocytsis* species within the mammalian malaria clade may indicate that the *Plasmodium* genus is paraphyletic (i.e., its phylogeny includes species of a different genus), a situation that would call for taxonomic revision. From Perkins and Schall, 2002.

transmitted to the human ancestral lineage from other, nonprimate hosts. These results are consistent with the diversity of physiological and epidemiological characteristics of these four *Plasmodium* species (Coatney *et al.*, 1971; López-Antuñano and Schumumis, 1993).

2. *Plasmodium falciparum* is more closely related to *P. reichenowi*, the chimpanzee parasite, than to any other *Plasmodium* species. On the basis of the *Csp* and *rRNA* genes, the time of divergence between these two *Plasmodium* species is estimated at 8–11 million years (My) ago, which is consistent with the time of divergence between the two host species, human and chimpanzee. (The divergence time of parasitic species is likely to predate the divergence of their host species, similarly as the divergence times of ancestral gene lineages are likely to predate the divergence of their species; alternative polymorphic states may become fixed in one or the other carrying species.) A parsimonious interpretation of this state of affairs is that *P. falciparum* is an ancient human parasite, associated with our ancestors since the divergence of the hominids from the great apes. The *Csp* tree confirms that *P. falciparum* is closely related to *P. reichenowi*, but distinct from it. Note that the cluster of the *P. falciparum* strains has a 92% bootstrap value, which is highly significant. (Figure 6.3 shows only five of the eight *falciparum* strains because others are identical to those shown.) The genetic distance between *falciparum* and *reichenowi* is 0.044 ± 0.002, five times as large as the average intraspecific distance among the eight *falciparum* strains, which is 0.009 ± 0.001 (Escalante *et al.*, 1995). The cytochrome-b tree (Figure 6.2) also confirms that *falciparum* is most closely related to *reichenowi* than to any other *Plasmodium* species, but different from it (Ayala *et al.*, 1998; Escalante *et al.*, 1998).

McCutchan *et al.* (1996) failed to separate unambiguously *P. falciparum* and *P. reichenowi* when they analyzed amino acid rather than nucleotide sequences. This ambiguity can be attributed to the difficulty of aligning, for several *Plasmodium* species, amino acid sequences that are quite different and variable in length (Escalante *et al.*, 1995; see also Rich *et al.*, 1997), with the consequence that only the more conserved amino acids can be reliably aligned. When the comparison is made between *P. reichenowi* and all available sequences of *P. falciparum*, the difference between the two species is unambiguous (see Rich *et al.*, 1997 for the distinct composition of the central *Csp* repeat region).

3. *Plasmodium malariae*, a human parasite and *P. brasilianum*, a New World monkey parasite, are genetically indistinguishable at the *Csp* gene. We infer that a lateral transfer between hosts has occurred in recent times, either from monkeys to humans or vice versa.

Moreover, *Plasmodium vivax* is genetically indistinguishable from *P. simium* at the *Csp* gene (and also at the *18SSUrRNA*; see Escalante *et al.*, 1997). *P. simium* is, like *P. brasilianum*, a parasite of New World monkeys. We infer, again, a recent lateral transfer between human and monkey hosts.

The average intraspecific distances for each of the three human parasites (and for *P. simium*; only one strain of *P. brasilianum* was investigated) are shown in Table 6.3. The Table also gives the genetic distance among the three human and two primate parasites. The genetic distance between *malariae* and *brasilianum* is 0.002 ± 0.002, not greater than the distance among the two *malariae,* or the four *vivax,* or the eight *falciparum* sequences available. Although *P. malariae* and *P. brasilianum* are isolated from very different hosts—*P. malariae* from humans, *P. brasilianum* from New World monkeys—the question arises whether they are different species, since they are genetically indistinguishable. *P. malariae* and *P. brasilianum* might be considered either two distinct species or a single species exhibiting "host polymorphism" (Escalante and Ayala, 1994) (i.e., able to parasitize more than one host species). But this is a question of taxonomy and convenience, rather than biologically substantive. Whether or not *malariae* and *brasilianum* are considered one or two distinct species, it would be the case that either *brasilianum* is a recent platyrrhine parasite acquired from humans, or *malariae* has only recently become a human parasite by a host–switch from New World monkeys. In the latter case, *malariae* would have become a human parasite within the last 15,000 years, after the first human colonization

Table 6.3. Average Genetic Distance Within and Between Various *Plasmodium* Species, Based on the *Csp* Gene

Species	Number of strains	Intraspecific	*malariae*	*vivax*	*simium*	*brasilianum*
falciparum	8	.009±.001	.697±.003	.581±.003	.837±.002	.687±.004
malariae	2	.004±.003		.517±.006	.513±.004	.002±.002
vivax	4	.004±.001			.004±.001	.517±.000
simium	2	.000±.000				.508±.187

of the Americas, and perhaps only within the last several hundred years, after the expansion of human populations that followed the European colonizations of South America.

3. Transfers Between Human and Monkey Hosts

The same issue, whether they should be considered only one or two species, arises with respect to *vivax* versus *simium*. Three considerations favor a lateral transfer from human to monkey hosts:

(i) *Plasmodium vivax* has a worldwide distribution, in contrast to the limited geographic range of *simium*, restricted to a few South American monkey species, *Alouatta fusca, Brachyteles arachnoides,* and *Ateles* sp. (Gysin, 1998). The counterpoint can be made, however, that humans are exceedingly mobile. Infected humans could readily have carried the parasite from South America to other world continents.

(ii) There are no records of malaria in South America (or elsewhere in the New World) before the arrival of the European colonizers within the last 500 years. This would be consistent with the interpretation that *P. vivax* (as well as *P. malariae* and *P. falciparum*) was introduced to the New World by the European colonizers and their African slaves. The weakness of this argument is that it consists of negative evidence, which is particularly unreliable when there are no extensive observations, experiments, or studies that would have likely manifested the presence of malaria in the New World before the year 1500, even if malaria had indeed been present.

(iii) Historical records suggest that nonmalignant malaria has occurred in the Old World for several thousand years. Chinese medical writings, dated 2700 BC, cuneiform clay tablets form Mesopotamia, dated about 2000 BC, the Ebere Egyptian Papyrus (ca. 1570 BC), and Vedic period Indian writings (1500–800 BC), mention severe periodic fevers, spleen enlargement and other symptoms suggestive of malaria (Sherman, 1998). Spleen enlargement and the malaria antigen have been detected in Egyptian mummies, some more than 3000 years old (Miller *et al.*, 1994; Sherman, 1998). Hippocrates' (460–370 BC) discussion of tertian and quartan fevers, "leaves little doubt that by the fifth century BC *Plasmodium malariae* and *P. vivax* were present in Greece" (Sherman, 1998, p. 3). If this interpretation is correct, the association of *malariae* and *vivax* with humans could not be attributed to a host–switch from monkeys to humans that would have occurred after the European colonizations of the Americas. This seems definitive evidence, so long as one accepts the interpretation that the fevers described by Hippocrates were indeed caused by the two particular species *P. vivax* and *P. malariae*.

Three considerations favor the alternative hypothesis, namely, that the host "invasion" has occurred from monkeys to humans:

(a) Humans are biologically (evolutionarily) more closely related to Old World monkeys (catharrhines) than to New World monkeys (platyrrhines). If lateral host–switch from humans to monkeys were likely, it would be more likely that the natural transfer would have been to our closer, rather than to our more remote relatives. This argument would be much weakened if the chimpanzee/gorilla parasite *P. rodhaini,* which is thought to be quite similar to *P. vivax*, or *P. schwetzi*, also a chimpanzee/gorilla parasite, which is similar to *P. malariae* (Gysin, 1998), were shown to be genetically identical (or very similar) to the corresponding human parasites, so that they might have been recently acquired by the apes from humans. We note that, according to Gysin (1998), "it has been shown" that *P. schwetzi* is "closely related" to *P. malariae*, but is "homologous" to *P. vivax*. This is a strange claim considering the considerable phylogenetic distance between *P. malariae* and *P. vivax* (see Figs. 6.2 and 6.3). We would point out, moreover, that *P. reichenowi*, and *P. falciparum*, which are "closely related" and "homologous" (Gysin, 1998) are evolutionarily as distant as their hosts, chimpanzees/gorillas and humans (see earlier section). The catharrhine parasite *P. inui*, widely thought to be closely related to *P. malariae*, has been shown to be quite different (Escalante *et al.*, 1998).

(b) Humans and their ancestors have been geographically associated with catarrhine monkeys for millions of years, but only for several thousand with platyrrhine monkeys. If the natural transfer from humans to monkeys were likely, it would have been much more likely that the transfer would have occurred to species with which humans have been in geographic association for a much longer period.

(c) *Plasmodium simium* is parasitic to several platyrrhine species. A lateral host transfer from humans to monkeys would require several host–switches, either from human to each monkey species, or from human to one monkey species and then from one to other monkey species, all in a short time interval (a few thousand, or even a few hundred years). This state of affairs is more extreme when we consider the case of *P. malariae* and *P. brasilianum*, since this parasite's hosts include numerous platyrrhine species (26 taxa are listed by Gysin, 1998, p. 420).

We have conjectured in the past (Escalante *et al.*, 1995; Ayala *et al.*, 1998), on the grounds of evolutionary parsimony, that the host–switch between *vivax* and *simium* and also between *malariae* and *brasilianum*—may have been from primates to humans so that *vivax* and *malariae* would have become human parasites only recently, perhaps only a few hundred years ago. The historical record (iii, see earlier section) is the strongest evidence against this conjecture. The matter can, in any case, be resolved by comparing the genetic diversity of the human and primate parasites. If the transfer has been from human to monkeys, the amount of genetic diversity in silent nucleotide sites (and other neutral polymorphisms) will be much greater in *P. vivax* than in *P. simium*, and in *P. malariae* than in *P. brasilianum* (including in each comparison the polymorphisms present in the several monkey host species). A transfer from monkey to humans would be evinced by much lesser polymorphism in the human than in the monkey parasites.

The genetic indistinguishability between *P. vivax* and *P. simium* has recently been confirmed by an investigation of 13 microsatellite loci and eight tandem-repeat (TR) loci, which includes 108 *P. vivax* individual samples broadly representative of the distribution of this parasite (Leclerc *et al.*, 2004). Microsatellite polymorphisms arise at high rates by replication slippage, yielding new alleles with different numbers of the repeating unit. The genetic near-identity between *P. vivax* and *P. simium* is evinced, first, by the fact that all 13 microsatellite loci could be amplified in *P. simium*. The number of microsatellite loci that could be amplified in any one of eight other Old World (catarrhyne) monkey parasites range from zero (*P. gonderi*) to ten (*P. cynomolgi*). Moreover, *P. simium* carries the same allele as *P. vivax* at the nine loci that are monomorphic in this species, the most common *P. vivax* allele at the three slightly polymorphic loci, and one of the *P. vivax* alleles at the only locus that is polymorphic in this species. A parallel situation occurs at the TR loci which evolve also rapidly. First, all eight TR loci could be amplified in *P. simium*, but only between zero and three in any of the other catarrhyne parasites. Second, *P. simium* alleles are identical to those of *P. vivax* at all loci but one, at which *P. simium* presents one private allele. This, however, is also the case for several local populations of *P. vivax*, which each display at least one private allele at one TR locus. A neighbor-joining tree based on TR genetic distances between populations includes *P. simium* within the *P. vivax* polymorphism (Leclerc *et al.*, 2004).

Lateral transmission of *Plasmodium* parasites from monkey hosts to humans is known for several species, including *P. simium* (Deane *et al.*, 1966), *P. brasilianum* (Contacos *et al.*, 1963), *P. cynomolgi* (Eyles *et al.*, 1960), *P. knowlesi* (Chin *et al.*, 1965) and, perhaps, *Plasmodium simiovale* (Qari *et al.*, 1993). Transmission from humans to monkeys can be accomplished experimentally (Chin *et al.*, 1965) and may also occur naturally (Collins, 1974). Among avian and reptilian malaria parasites, host–shifts have been a common occurrence (Bensch *et al.*, 2000; Ricklefs and Fallon, 2002).

The direction of host transfer between *P. vivax* and *P. simium* can be settled genetically. If the transfer has occurred from platyrrhine to human, *P. simium* is expected to have greater nucleotide polymorphism at neutral sites than *P. vivax*. Ascertaining this will require the investigation of numerous independent samples of *P. simium*, which are not currently available. Moreover, the issue may not simply be resolved, because the genetic impoverishment of *P. vivax* may be due to a recent demographic (or selective) sweep, which for now is postulated as the likely explanation for this impoverishment (Leclerc *et al.*, 2004; Rich 2004).

4. Population Structure of *Plasmodium falciparum*

As shown earlier, the lineage of *P. falciparum* has been associated with the human lineage throughout the evolution of the hominids (i.e., since the separation of the human and chimpanzee lineages), 6–8 million years (My) ago. The question we now raise is whether the population dynamics of *P. falciparum* has or not followed that of its hominid hosts. A possibility, for example, is that *P. falciparum* may have been restricted to some small locality from which it might have spread throughout other human populations as a consequence of increased virulence, environmental changes, or some other factor.

One way to approach this question is to investigate the distribution of genetic polymorphisms in *P. falciparum* populations. Numerous epidemiological studies have indicated that populations of *P. falciparum* are remarkably variable. Extensive genetic polymorphisms have been identified with respect to antigenic determinants, drug resistance, allozymes, and chromosome sizes (e.g., Sinnis and Wellems, 1988; Creasy *et al.*, 1990; Kemp and Cowan, 1990; McConkey *et al.*, 1990; Hughes and Hughes, 1995; Babiker and Walliker, 1997). Antigenic and drug resistance polymorphisms respond to natural selection, which is most effective in large populations—millions of humans are infected by *P. falciparum* and one single patient may harbor 10^{10} parasites (McConkey *et al.*, 1990). The replacement of one allele by another, or the rise of polymorphism with two or more alleles at high frequency may occur even in one generation. If the selection pressure is strong enough, all individuals exposed to the selective agent may die, except those carrying a resistant mutation. With populations as large as those of *P. falciparum*, any particular mutation is expected to arise in any one generation; and the same mutation may arise—and rise to large frequency—independently in separate populations. On the contrary, silent (i.e., synonymous) nucleotide polymorphisms are often adaptively neutral (or very nearly so) and not directly subject to natural selection. Thus, silent nucleotide polymorphisms reflect the mutation rate and the time elapsed since their divergence from a common ancestor. The population structure of *P. falciparum* is, consequently, best investigated by examining the incidence of synonymous polymorphisms.

The coalescence theory of population genetics assumes that the allele sequences of any given gene present in populations of an organism can be genealogically traced back to a single ancestral sequence ("cenancestor"). If one ignores the possibility of multiple hits (which is reasonable, so long as the sequences are not extremely polymorphic), the number of neutral polymorphisms observed in a sample of multiple strains will be a function of the neutral mutation rate, the time elapsed, and the number of lineages examined (and follow a Poisson distribution). If the neutral mutation rate can be established, the time elapsed since the cenancestor is simply determined by dividing the number of polymorphisms observed by the mutation rate times the number of neutral nucleotide sites observed in the full sample (Appendix).

Table 6.4 summarizes the polymorphisms that we found in a sample of 10 genes for which several sequences were available in DNA data banks (Rich *et al.*, 1998). The gene sequences analyzed derive form isolates of *P. falciparum* representative of the global malaria endemic regions. The *Dhfr* and *Ts* genes are found directly adjacent to one another on the parasite's fourth chromosome and encode the bifunctional dihydrofolate reductase–thymidylate synthetase (DHFR–TS) domain. Certain mutations in the *Dhfr* gene have been widely associated with *P. falciparum* resistance to antifolate drugs, including pyrimethamine. Two other genes in Table 6.4 have been implicated with drug-resistant phenotypes of *P. falciparum*: the gene coding for dihydropteroate synthetase (*Dhps*) and the gene for multidrug resistance (*Mdr1*). The circumsporozoite protein (encoded by *Csp1*) is antigenic, and the rhoptry-associated protein (encoded by *Rap1*) may also be immunogenic. The other four genes in Table 6.4 are not known to be immunogenic or associated with resistance to any antimalarial drug currently in use. They code for calmodulin (*Calm*), glucose-6-phosphate dehydrogenase (*G6pd*), heat-

Table 6.4. Polymorphisms in 10 Gene Loci of *Plasmodium falciparum*

Gene	Chromosome location	Length (bp)	Sample size	Polymorphic sites		Number of synonymous sites	
				nonsynonymous	synonymous	Fourfold	Twofold
Dhfr	4	609	32	4	0	2144	4128
Ts	4	1215	10	0	0	1250	2640
Dhps	8	1269	12	5	0	1536	2724
Mdr1	5	4758	3	1	0	1350	2088
Rap1	—	2349	9	8	0	1092	1668
Calm	14	441	7	0	0	364	602
G6pd	14	2205	3	9	0	726	1404
Hsp86	7	2241	2	0	0	532	910
Tpi	—	597	2	0	0	180	262
Csp1 5′ end	3	387	25	7	0	688	2010
Csp1 3′ end	3	378	25	17	0	1050	1625
Total	—	—	—	51	0	10,912	20,061

shock protein 86 (*Hsp86*), and triose phosphate isomerase (*Tpi*). Six of the ten loci exhibit amino acid polymorphisms, including the drug-resistance genes *Dhfr*, *Dhps,* and *Mdr-1*, as well as the antigenic *Cps* and *Rap1*. The significant result is that no silent polymorphisms are observed in any of the 10 genes. An independent study of ten gene loci, most encoding antigenic determinants, has shown a similar scarcity of silent polymorphisms (Escalante *et al.*, 1998).

5. Malaria's Eve Hypothesis

Estimating time of divergence from neutral polymorphism requires that the neutral mutation rate (of third position in synonymous codons) be known. We have estimated the neutral mutation rate by comparing the *P. falciparum* gene sequences with *P. reichenowi* (the chimpanzee parasite) and also with a set of rodent *Plasmodium* parasites (Rich et al., 1998). The number of neutral polymorphisms in Table 6.4 is zero and, thus, at face value, the time elapsed since the cenancestor (t) would be zero, although it would become positive as soon as some neutral polymorphisms are observed. In any case, t is expected to follow a Poisson distribution, which allows calculating the upper confidence limit for the time since the cenancestor. As shown in Table 6.5, the 95% upper confidence level is between 25,000–50,000 years ago, depending on which mutation rate estimate is used; the 50% upper confidence limit is between 6000–13,000 years. We have referred to this conclusion, that the world expansion of *P. falciparum* is recent, as the Malaria's Eve hypothesis.

Table 6.5. Estimated Upper-boundary Times (t_{95} and t_{50}, in Years) to the Cenancestor of the World Populations of *P. falciparum*

	Mutation rate × 10^{-9}			
Assumption	Fourfold	Twofold	t_{95}	t_{50}
Plasmodium radiation				
55 My	7.12	2.22	24,511	5,670
129 My	3.03	0.95	57,481	13,296
falciparum-reichenowi				
5 My	3.78	1.86	38,136	8,821
7 My	2.70	1.33	53,363	12,342

Mutation rates are estimated based on two sets of assumptions, concerning either the origin of the *Plasmodium* genus or the time of divergence between *P. falciparum* and *P. reichenowi*. The *P. falciparum* cenancestor lived more recently than 24,511 or 57,481 years ago, with a 95% probability; and more recently than 5670 or 13,296 years with a probability of 50%.

In the few years since we first proposed the Malaria's Eve hypothesis (1998), the issue has been subject to a contentious debate. Our initial conclusion was based on sequences that were then available from GenBank, and the only criteria for inclusion of genes in our dataset was that they had to be void of repetitive DNA sequences and show no evidence of being under positive selection impacting synonymous codon substitutions. In 1998, the amount of sequence data available for the species was rather limited, but since that time the dataset has grown enormously, including the complete genome sequence of *P. falciparum* published in 2002 (Gardner *et al.*, 2002).

One of the new studies entails a large-scale sequencing survey of 25 introns, located on the second chromosome, from eight *P. falciparum* isolates collected among global sites (Volkman *et al.*, 2001). The findings of this study confirm our previous result: there is an extreme scarcity of silent-site polymorphism among extant populations of *P. falciparum*. Among some 32,000 nucleotide sites examined, Volkman *et al.* (2001) found only three silent single-nucleotide polymorphisms (SNPs). Combining their data with ours, these authors estimated that the age of Malaria's Eve was somewhere between 3200 and 7700 years, depending on the calibration of the substitution clock.

Conway *et al.* (2000) have presented further evidence in support of Malaria's Eve, based on analysis of the *P. falciparum* mitochondrial genome. They examined the entire mitochondrial DNA (mtDNA) sequence of four *P. falciparum* isolates originating from Africa, Brazil, and two from Thailand, as well as the chimpanzee parasite, *P. reichenowi*. Alignment of the four complete mtDNA sequences (5965 bp) showed that 139 sites contain fixed differences between *falciparum* and *reichenowi*, whereas only four sites are polymorphic within *falciparum*. The corresponding estimates of divergence (*K*, between *P. reichenowi* and *P. falciparum*) and diversity (π, within *P. falciparum* strains), are 0.1201 and 0.0004, respectively (i.e., divergence in *mtDNA* sequence between the two species is 300-fold greater than the diversity within the global *P. falciparum* population). If we use the *rDNA*-derived estimate of 8 million years as divergence time between *P. falciparum* and *P. reichenowi*, then the estimated origin of the *P. falciparum* mtDNA lineages is 26,667 years (i.e., 8 million/300), which corresponds quite well with our estimate based on 10 nuclear genes (Rich *et al.*, 1998). In a subsequent survey of a total of 104 isolates from Africa ($n = 73$), Southeast Asia ($n = 11$), and South America ($n = 20$), Conway *et al.* (2000) determined that the extant global population of *P. falciparum* is derived from three mitochondrial lineages that started in Africa, and migrated subsequently (and independently) to South America and Southeast Asia. Each mitochondrial lineage is identified by a unique arrangement of the four polymorphic *mtDNA* nucleotide sites.

More recently, the complete 6-kb mitochondrial genome has been sequenced from 100 geographically representative isolates, (Joy *et al*, 2003). The sequences support a rapid expansion of *P. falciparum* in Africa, starting approximately 10,000 years ago. The data indicate, however, that some lineages are five to ten times more ancient and that *P. falciparum* populations have existed in Southeast Asia and South America perhaps earlier than 50,000 years ago. This is unexpected, because if *falciparum* originated in Africa, the oldest polymorphisms should occur in that continent, as is the case for human mitochondrial lineages. Moreover, the presence of *P. falciparum* in South America more than 50,000 years ago is most unlikely, since humans colonized America 15,000–18,000 years ago (or, less likely, around 30,000 years ago) (Cavalli-Sforza *et al.*, 1994) and, moreover, it is generally accepted that *falciparum* malaria was introduced in America by the slave trade (Watts, 1997). Rather, it seems likely that the high differentiation of some mitochondrial sequences may be a consequence of the large variation expected in a sample of nonrecombinant neutral sequences, as is the case for the mitochondrial genome, or a consequence of natural selection increasing the frequency of some favorable sequences, which then would appear to be much older than they actually are. Some mitochondrial genes are likely to be under selection, such as the gene encoding cytochrome b, which seems to underlie susceptibility to some antibiotics (Vaidya *et al.*, 1993).

A recent investigation of 20 protein-coding genes has confirmed the low level of synonymous polymorphisms found in earlier studies

(Table 6.4; see also Escalante *et al.*, 1998; Rich *et al.*, 1998). Many of the 20 loci were chosen so as to be different from those previously analyzed, and are located on 11 different chromosomes. The 20 genes were sequenced in 5–7 reference isolates. Among the 22,611 nucleotides sequenced for each strain, there were 21 nonsynonymous polymorphisms, but only one synonymous polymorphism (Table 6.2 in Hartl, 2004), strongly supporting the Malaria's Eve hypothesis.

6. Malaria's Eve Counterarguments

Arguments against the Malaria's Eve hypothesis come in two forms. The first argument is that the loci chosen in the studies supporting the recent world expansion of *P. falciparum* are a biased sample and do not reflect the levels of polymorphism in the genome as a whole. The second argument concedes that nucleotide polymorphisms are scarce and proposes that this is not attributable to recent origins, but rather reflects strong selection pressure against the occurrence of synonymous substitutions.

Some studies have estimated that antigenic polymorphisms in *P. falciparum* may be 40 or more million years old, older than the origin of the hominids (Hughes, 1993; Hughes and Hughes, 1995). An analysis of nucleotide sequences for 23 gene loci has concluded that the cenancestor of extant populations of *P. falciparum* must be 300,000–400,000 years old (Hughes and Verra, 2001). As we have previously explained, these inferences ignore natural selection promoting rapid evolution of antigenic determinants; rely on erroneous sequence alignments that fail to recognize the presence of repetitive sequences; cannot account for large excess of nonsynonymous over synonymous substitutions; and depend on poorly scrutinized sequences obtained from data banks (Rich and Ayala, 1999, 2000, 2003). Some alleged polymorphisms are sequencing errors since they come from a single gene derived from a single clone, but sequenced in different laboratories (see Rich and Ayala, 2003; Hartl, 2004). Similar problems plague the survey by Mu *et al.* (2002) of more than 200 kb from the complete chromosome 3 and their conclusion that *P. falciparum* has maintained large populations for at least 300,000 years (Ayala and Rich, 2003; Hartl, 2004).

The argument that selection pressure against synonymous substitutions, rather than a recent cenancestor, can account for the scarcity of synonymous polymorphisms has also been answered in detail (Rich and Ayala, 1998, 2003). This argument is largely based on the predominance of AT pairs over GC pairs in the genome of *P. falciparum*. Indeed, the AT content of *P. falciparum* is 71.7% overall, and 83.6% in the third position

(Nakamura *et al.*, 1997). In response, it may suffice here to say that: (1) AT excess lowers the rate of synonymous substitution but does not altogether eliminate it; (2) in fourfold redundant codons, the bias favors codons ending in A or T but it does not impact A↔T mutations; (3) other *Plasmodium* species are also AT rich and should have the same mutational constraints as *P. falciparum*, yet they exhibit abundant synonymous intraspecific polymorphisms as well as interspecific differentiation; (4) comparisons between *P. falciparum* and *P. reichenowi* at five genes for which data are available in both species yield high numbers of synonymous substitution (average $K_s = 0.072$ versus $K_n = 0.046$ for nonsynonymous substitutions) (Rich and Ayala, 1998).

Beyond genetic inference, other considerations support the Malaria's Eve hypothesis. Sherman (1998) has noted the late introduction and low incidence of *falciparum* malaria in the Mediterranean region. Hippocrates (460–370 BC) describes quartan and tertian fevers, but there is no mention of severe malignant tertian fevers, which suggests that *P. falciparum* infections did not yet occur in classical Greece, as recently as 2400 years ago.

The expansion of *P. falciparum* across the globe after the Neolithic revolution, perhaps about 5,000–6,000 years ago, starting from a highly restricted geographic location, probably in tropical Africa, may have been made possible by (1) changes in human societies, (2) genetic changes in the host–parasite-vector association that have altered their compatibility, and (3) climatic changes that entailed demographic changes (migration, density, etc.) in the human host, the mosquito vectors, and/or the parasite.

One factor may have been changes in human living patterns, particularly the development of agricultural societies and urban centers that increased human population density (Livingston, 1958; Weisenfeld, 1967; De Zulueta *et al.*, 1973; de Zulueta, 1994; Coluzzi, 1997, 1999; Sherman, 1998). Genetic changes that have increased the affinity within the parasite–vector–host system are also a possible explanation for a recent expansion, not mutually exclusive with the previous one. Coluzzi (1997, 1999) has cogently argued that the recent worldwide distribution of *P. falciparum* has come about, in part, as a consequence of a recent dramatic rise in vectorial capacity due to repeated speciation events in Africa of the most anthropophilic members of the species complexes of the *Anopheles gambiae* and *A. funestus* mosquito vectors. Biological processes implied by this account (2, above) may have been associated with, and even be dependent on the onset of agricultural societies in Africa (1, above) and climatic changes (3, above), specifically gradual increase in ambient temperatures after the Würm glaciation, so that about 6000 years ago climatic conditions in the Mediterranean region and the Middle East made the spread of *P. falciparum* and its vectors beyond tropical Africa possible (De Zulueta *et al.*, 1973; de Zulueta, 1994; Coluzzi, 1997, 1999). Once demographic and climatic conditions became suitable for propagation of *P. falciparum*, natural selection would have facilitated evolution of *Anopheles* species that were highly anthropophilic and effective *falciparum* vectors (De Zulueta *et al.* 1973; Coluzzi 1997, 1999).

The Malaria's Eve hypothesis is consistent with the evolutionary history of genetically determined immunity factors that confer resistance to *P. falciparum*. A glucose-6-phosphate dehydrogenase (*G6pd*) genetic deficiency occurs at high frequency in areas of malaria endemicity, particularly in subSaharan Africa. Tishkoff *et al.* (2001) have concluded that the *G6pd* mutants are only 3330 years old (95% confidence interval, 1600–6640 years), indicating that *P. falciparum* has only recently become a disease burden to humans. It has been known for decades that hemoglobin S (Hbs; βGglu→Val) heterozygosity confers protection against severe malaria (Allison, 1964; Hill and Weatherall, 1998). Current evidence suggests that the sickle cell mutation has arisen at least twice, once in India or the Middle East and once in Africa, although it is possible that the mutation may have arisen more than once in Africa (Hill and Weatherall, 1998). Analyses of the β globin gene haplotypes associated with a sickle-cell mutation in Africa suggest that it is recent in origin, perhaps no more than 2000 years (Currat *et al.*, 2002; Weatherall, 2004). A hemoglobin C (Hbc; β6Glu→Lys) mutation seems to have arisen in Africa equally recently, or even more recently. HbC is associated with a 29% reduction in risk of clinical malaria in heterozygotes (HbCA) and of 93% reduction in HbCC homozygotes, which exhibit limited pathology compared to the severely disadvantaged HbSS (Modiano *et al.*, 2001).

Appendix

If a population grows to a large size after a bottleneck, it is reasonable to assume that the genealogy of a sample of multiple strains collected from widely distributed localities would be a star-like phylogeny with the common ancestor at the vertex of the star, as in Figure 6.4 (Slatkin and Hudson, 1991). Under this assumption, and ignoring the possibility of multiple hits at individual sites, the number of neutral polymorphisms that we observe in a sample of multiple strains will have a Poisson distribution with a mean that depends on the neutral mutation rate, the time elapsed, and the number of lineages examined. The expected number of polymorphisms is

$$\lambda = \mu_a t \sum n_i l_i + \mu_b t \sum n_i m_i,$$

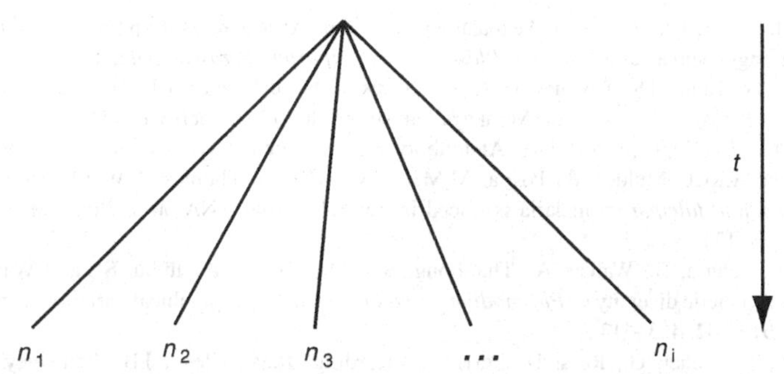

Figure 6.4. Schematic representation of a star phylogeny with the most recent common ancestor (MRCA) at the apex. The scale of *t* represents the time elapsed since this MRCA gave rise to all its descendents in the extant population (n_1, n_2, n_3, ... n_i).

where μ_a and μ_b are the neutral mutation rates at the third position of fourfold and twofold degenerate codons, respectively; *t* is the time since the bottleneck; n_i the number of lineages sampled at the *i*th locus; and l_i and m_i, respectively, the number of fourfold and twofold synonymous sites examined at the *i*th locus. This expression suggests an estimator of the time of the bottleneck, obtained by solving for *t* and replacing λ (the *expected* number of polymorphisms) by *S*, the *observed* number of polymorphisms:

$$\hat{t} = \frac{S}{\mu_a \sum n_i l_i + \mu_b \sum n_i m_i,}.$$

References

Allison, A.C. (1964). Polymorphism and natural selection in human populations, *Cold Spring Harbor Symp. Quant. Biol.*, **29,** 137–149.

Ayala, F., Escalante, A., Lal, A., and Rich, S. (1998). Evolutionary relationships of human malarias. In I.W. Sherman. (ed.), *Malaria: Parasite Biology, Pathogenesis, and Protection.* American Society of Microbiology, Washington, DC. pp. 285–300.

Ayala, F.J., Escalante, A.A., and Rich, S.M. (1999). Evolution of plasmodium and the recent origin of the world populations of *Plasmodium falciparum. Parassitologia*, **41,** 55–68.

Babiker, H. and Walliker, D. (1997). Current views on the population structure of *Plasmodium falciparum*: Implications for control. *Parasitol. Today*, **13,** 262–267.

Barta, J.R. (1989). Phylogenetic analysis of the class sporozoea (phylum Apicomplexa Levine, 1970): Evidence for the independent evolution of heteroxenous life cycles. *J. Parasitol.*, **75,** 195–206.

Bensch, S., Stjernman, M., Hasselquist, D., Ostman, O., Hansson, B., Westerdahl, H., and Pinheiro, R.T. (2000). Host specificity in avian blood parasites: A study of plasmodium and haemoproteus mitochondrial DNA amplified from birds. *Proc. Royal Soc. London Ser. B-Biol. Sci.*, **267,** 1583–1589.

Cavalli-Sforza, L.L., Menozzi, P., and Piazza, A. (1994). *The History and Geography of Human Genes.* Princeton University Press, Princeton, NJ.

Chin, W., Contacos, P.G., Coatney, G.R., and Kimball, H.R. (1965). A naturally acquired quotidian-type malaria in man transferable to monkeys. *Science*, **149,** 865.

Coatney, R.G., Collins, W.E., Warren, M., and Contacos, P.G. (1971). *The Primate Malarias.* U.S. Government Printing Office, Washington DC.

Coluzzi, M. (1997). *Evoluzione Biologica i Grandi Problemi della Biología.* Accademia dei Lincei, Rome, pp. 263–285.

Coluzzi, M. (1999). The clay feet of the malaria giant and its African roots: Hypotheses and inferences about origin, spread and control of *Plasmodium falciparum. Parassitologia*, **41,** 277–283.

Contacos, P.G., Lunn, J.S., Coatney, G.R., Kilpatrick, J.W., and Jones, F.E. (1963). Quartan-Type Malaria Parasite of New World Monkeys Transmissible to Man. *Science*, **142**, 676.

Conway, D.J., Fanello, C., Lloyd, J.M., Al-Joubori, B.M., Baloch, A.H., Somanath, S.D., Roper, C., Oduola, A.M.J., Mulder, B., Povoa, M.M., Singh, B., and Thomas, A.W. (2000). Origin of *Plasmodium falciparum* malaria is traced by mitochondrial DNA. *Mol. Biochem. Parasitol.*, **111,** 163–171.

Creasey, A., Fenton, B., Walker, A., Thaithong, S., Oliveira, S., Mutambu, S., and Walliker, D. (1990). Genetic diversity of *Plasmodium falciparum* shows geographical variation. *Am. J. Trop. Med. Hyg.*, **42,** 403–413.

Currat, M., Trabuchet, G., Rees, D., Perrin, P., Harding, R.M., Clegg, J.B., Langaney, A., and Excoffier, L. (2002). Molecular analysis of the β-globin gene cluster in the Niokholo Mandenka population reveals a recent origin of the β^S Senegal mutation. *Am.J. Hum. Gen.*, **70,** 207–223.

Deane, L.M., Deane, M.P., and Ferreira, N.J. (1966). *Trans. R. Soc. Trop. Med. Hyg.* **60,** 563–564.

de Zulueta, J., (1994). Malaria and ecosystems: From prehistory to posteradication. *Parassitologia*, **36,** 7–15.

De Zulueta, J., Blazquez, J., and Maruto, J.F. (1973). Entomological aspects of receptivity to malaria in the region of Navalmoral of Mata. *Rev. Sanid. Hig. Publica.* (Madr) **47,** 853–870.

Eyles, D.E., Coatney, G.R. and Getz, M.E. (1960). Vivax-type malaria parasite of macaques transmissible to man. *Science,* **131,** 1812–1813.

Escalante, A.A. and Ayala, F J. (1994). Phylogeny of the malarial genus *Plasmodium* derived from rRNA gene sequences. *Proc. Natl. Acad. Sci. USA.*, **91,** 11373–11377.

Escalante, A.A. and Ayala, F.J., (1995). Evolutionary origin of *Plasmodium* and other Apicomplexa based on rRNA genes. *Proc. Natl. Acad. Sci. USA*, **92,** 5793–5797.

Escalante, A.A., Barrio, E. and Ayala, F.J. (1995). Evolutionary origin of human and primate malarias: evidence from the circumsporozoite protein gene. *Mol. Biol. Evol.*, **12,** 616–626.

Escalante, A.A., Goldman, I.F., De Rijk, P., De Wachter, R., Collins, W.E., Qari, S.H. and Lal, A.A. (1997). Phylogenetic study of the genus *Plasmodium* based on the secondary structure-based alignment of the Small Subunit ribosomal RNA. *Mol. Biochem. Parasitol.*, **90,** 317–321.

Escalante, A.A., Lal, A.A. and Ayala, F.J. (1998). Genetic polymorphism and natural selection in the malaria parasite *plasmodium falciparum. Genetics*, **149,** 189–202.

Fast, N.M., Xue, L., Bingham, S. and Keeling, P.J. (2002). Re-examining alveolate evolution using multiple protein molecular phylogenies. *J. Eukaryot. Microbiol.*, **49,** 30–37.

Felsenstein, J. (1985). Confidence limits on phylogenies: an approach using bootstrap, *Evolution*, **39,** 783–791.

Gardner, M.J., Hall, N., Fung, E., White, O., Berriman, M., Hyman, R.W., Carlton, J.M., Pain, A., Nelson, K.E., Bowman, S., Paulsen, I.T., James, K., Eisen, J.A., Rutherford, K., Salzberg, S.L., Craig, A., Kyes, S., Chan, M.S., Nene, V., Shallom, S.J., Suh, B., Peterson, J., Angiuoli, S., Pertea, M., Allen, J., Selengut, J., Haft, D., Mather, M.W., Vaidya, A.B., Martin, D.M.A., Fairlamb, A.H., Fraunholz, M.J., Roos, D.S., Ralph, S.A., McFadden, G.I., Cummings, L.M., Subramanian, G.M., Mungall, C., Venter, J.C., Carucci, D.J., Hoffman, S.L., Newbold, C., Davis, R.W., Fraser, C.M., and Barrell, B. (2002). Genome sequence of the human malaria parasite *Plasmodium falciparum. Nature* **419,** 498–511.

Garnham, P.C.C. (1966). *Malaria Parasites and Other Haemosporidia*. Blackwell Scientific Publications, Oxford, UK.

Gysin, J. (1998). Animal models: Primates. In I.W. Sherman (ed), *Malaria: Parasite Biology, Pathogenesis, and Protection*, ASM Press, Washington, DC. pp. 419–441.

Hartl, D.L. (2004). The origin of malaria: mixed messages from genetic diversity. *Nat. Rev. Microbiol.*, **2,** 15–22.

Hill, A.V.S. and Weatherall, D.J. (1998). Host genetic factors in resistance to malaria, In I. W. Sherman (ed), *Malaria: Parasite Biology, Pathogenesis, and Protection*. American Society of Microbiology, Washington, D.C., pp. 445–455.

Huff, C.G. (1938). Studies on the evolution of some disease-producing organisms. *Q. Rev. Biol.* **13,** 196–206.

Hughes, A.L. (1993). Coevolution of immunogenic proteins of *Plasmodium falciparum* and the host's immune system, In N. Takahata, and A.G. Clark (eds), *Mechanisms of Molecular Evolution*, Sinauer Assoc., Sunderland, Mass, U. S. A., pp. 109–127.

Hughes, A.L., and Hughes, M.K. (1995). Natural Selection on *Plasmodium* surface proteins. *Mol. Biochem. Parasitol.*, **71,** 99–113.

Hughes, A.L. and Verra, F. (2001). Very large long-term effective population size in the virulent human malaria parasite *Plasmodium falciparum. Proc. R. Soc. Lond. B. Biol. Sci.*, **268,** 1855–1860.

Joy, D.A., Feng, X., Mu, J., Furuya, T., Chotivanich, K., Krettli, A.U., Ho, M., Wang, A., White, N.J., Suh, E., Beerli, P. and Su, X.Z. (2003). Single-nucleotide polymorphisms and genome diversity in *Plasmodium vivax, Science*, **300,** 318–321.

Kemp, D.J. and Cowman, A.F. (1990). Genetic diversity of *Plasmodium falciparum. Adv. Parasitol.*, **29,** 75–133.

Kimsey, R.B. (1992). Host association and the capacity of sand flies as vectors of lizard malaria in Panama. *Int. J. Parasitol.*, **22,** 657–664.

Leclerc, M.C., Durand, P., Gauthier, C., Patot, S., Billotte, N., Menegon, M., Severini, C., Ayala, F.J. and Renaud, F. (2004). Meager genetic variability of the human malaria agent *Plasmodium vivax. Proc. Natl. Acad. Sci. USA*, **101,** 14455–14460.

Livingston, F.B. (1958). Anthropological Implications of sickle cell gene distribution in West Africa. *American Anthropologist* **60,** 533–560.

López-Antuñano, F. and Schumunis, F.A. (1993). Plasmodia of humans, In J. P. Kreier, (ed), *Parasitic Protozoa*, 2nd edn., vol. **5**, Academic Press Inc., New York, pp. 135–265.

Manwell, R. (1955). Some evolutionary possibilities in the history of the malaria parasites. *Indian J. Malariol.*, **9**,247–253.

Margulis, L., McKhann, H., and Olendzenski, L. (1993). *Illustrated Guide of Protoctista*. Jones and Bartlett, Boston.

McConkey, G.A., Waters, A.P. and McCutchan, T.F. (1990). The generation of genetic diversity in malarial parasites. *Annu. Rev. Microbiol.*, **44,** 479–498.

McCutchan, T.F., Kissinger, J.C., Touray, M.G., Rogers, M.J., Li, J., Sullivan, M., Braga, E.M., Kretli, A.U. and Miller, L. (1996). Comparison of circumsporozoite proteins from avian and mammalian malaria: *Proc. Natl. Acad. Sci. USA* **93,** 11889–11894.

Miller, R.L., Ikram, S., Armelagos, G.J., Walker, R., Harer, W.B., Shiff, C.J., Baggett, D., Carrigan, M. and Maret, S.M. (1994). Diagnosis of *Plasmodium falciparum* infections in mummies using the rapid manual ParaSight™-Ftest. *Trans. R. Soc. Trop. Med. Hyg.*, **88,** 31–32.

Modiano, D., Luoni, G., Sirima, B.S., Simporé, J., Verra, F., Konaté, A., Rastrelli, E., Olivieri, A., Calissano, C., Paganotti, G.M., D'Urbano, L., Sanou, I., Sawadogo, A., Mediano, G. and Coluzzi, M. (2001). Haemoglobin C protects against clinical *Plasmodium falciparum* malaria. *Nature*, **414,** 305–308.

Mu, J., Duan, J., Makova, K.D., Joy, D.A., Huynh, C.Q., Branch, O.H., Li, W.-H. and Su, X.-Z (2002). Chromosome-wide SNPs reveal an ancient origin for *Plasmodium falciparum. Nature*, **418,** 323–326.

Nakamura, Y., Gojobori, T. and Ikemura, T. (1997). Codon usage tabulated from the international DNA sequence databases. *Nuc. Acids Res.*, **25,** 244–245.

Perkins, S.L. and Schall, J.J. (2002). A molecular phylogeny of malarial parasites recovered from cytochrome-b gene sequences. *J. Parasitol.*, **88,** 972–978.

Qari, S.H., Shi, Y.-P., Povoa, M.M., Alpers, M.P., Deloron, P., Murphy, G.S., Harjosuwarno, S. and Lal, A.A. (1993). *J. Infec. Dis.*, **168,** 485–1489.

Qari, S.H., Shi, Y.P., Pieniazek, N.J., Collins, W.E. and Lal, A.A. (1996). Phylogenetic relationship among the malaria parasites based on small subunit rRNA gene sequences: Monophyletic nature of the human malaria parasite. *Plasmodium falciparum. Mol. Phylogenet. Evol.*, **6,** 157–165.

Rich, S.M. (2004). The unpredictable past of *Plasmodium vivax* revealed in its genome. *Proc. Natl. Acad. Sci. USA*, **101,** 15547–15548.

Rich, S.M. and Ayala, F.J. (1998). The recent origin of allelic variation in antigenic determinants of *Plasmodium falciparum*. *Genetics,* **150,** 515–517.

Rich, S.M. and Ayala, F. J. (1999). Circumsporozoite polymorphism, silent mutations and the evolution of Plasmodium *falciparum*. Reply. *Parasitol. Today*, **15,** 39–40.

Rich, S.M. and Ayala, F.J. (2000). Population structure and recent evolution of *Plasmodium falciparum*. *Proc. Natl. Acad. Sci. USA*, **97,** 6994–7001.

Rich, S.M. and Ayala. F.J. (2003). Progress in malaria research: the case for phylogenetics. In D.T.J. Littlewood (ed), *Advances in Parasitology: The Evolution of Parasitism a Phylogenetic Perspective*, vol. **54.** Elsevier/Academic, Amsterdam. pp. 255–280.

Rich, S.M., Hudson, R.R., and Ayala, F.J. (1997). *Plasmodium falciparum* antigenic diversity: Evidence of clonal population structure. *Proc. Natl. Acad. Sci. USA*, **94,** 13040–13045.

Rich, S.M., Licht, M.C., Hudson, R.R., and Ayala, F.J. (1998). Malaria's Eve: Evidence of a recent bottleneck in the global *Plasmodium falciparum* population. *Proc. Natl. Acad. Sci. USA,* **95,** 4425–4430.

Ricklefs, R.E. and Fallon, S.M. (2002). Diversification and host–switching in avian malaria parasites. *Proc. Royal Soc. London B.*, **269,** 885–892.

Saitou, N. and Nei, M. (1987). The neighbor-joining method: A new method for reconstructing phylogenetic trees. *Mol. Biol. Evol.*, **4,** 406–425.

Sherman, I.W. (1998). A brief history of malaria and the discovery of the parasite's life cycle. In: I.W. Sherman (ed), *Malaria: Parasite Biology, Pathogenesis, and Protection.* American Society of Microbiology, Washington, DC. pp. 3–10.

Sinnis, P. and Wellems, T.E. (1988). Long range restriction maps of *Plasmodium falciparum* chromosomes: Crossing over and size variation in geographically distant isolates. *Genomics*, **3,** 287–295.

Slatkin, M. and Hudson, R.R. (1991). Pairwise comparisons of mitochondrial DNA sequences in stable and exponentially growing populations. *Genetics*, **129,** 555–562.

Tamura, K. (1992). Estimation of the number of nucleotide substitutions when there are strong transition–transversion and G+C content biases. *Mol. Biol. Evol.*, **9,** 678–687.

Tishkoff, S.A., Varkonyi, R., Cahinhinan, N., Abbes, S., Argyropoulos, G., Destro-Bisol, G., Drousiotou, A., Dangerfield, B., Lefranc, G., Loiselet, J., Piro, A., Stoneking, M., Tagarelli, A., Tagarelli, G., Touma, E.H., Williams, S.M., and Clark, A.G. (2001). Haplotype diversity and linkage disequilibrium at human G6PD: Recent origin of alleles that confer malarial resistance. *Science,* **293,** 455–462.

Vaidya, A.B., Lashgari, M.S., Pologe, L.G., and Morrissey, J. (1993). Structural features of *Plasmodium* cytochrome-b that may underlie susceptibility to 8-aminoquinolines and hydroxynaphthoquinones. *Mol. Biochem. Parasitol.*, **58,** 33–42.

Volkman, S.K., Barry, A.E., Lyons, E.J., Nielsen, K.M., Thomas, S.M., Choi, M., Thakore, S.S., Day, K.P., Wirth, D.J., and Hartl, D.L. (2001). Recent origin of *Plasmodium falciparum* from a single progenitor. *Science*, **293,** 482–484.

Watts, S. (1997). *Epidemics and History: Disease, Power and Imperialism.* Yale University Press, New Haven, CT.

Weatherall, D.J. (2004). J.B.S. Haldane and the malaria hypothesis. In K.R. Dronamraju (ed), *Infectious Disease and Host-Pathogen Evolution.* Cambridge University Press, Cambridge. pp. 18–36.

Weisenfeld, S.L. (1967). Sickle cell trait in human biological and cultural evolution. Development of agriculture causing increased malaria is bound to gene pool changes causing malaria reduction. *Science* **157,** 1134–1140.

Vector Genetics in Malaria Control

V.P. Sharma

Abstract

Anopheles culicifacies is the major vector of rural malaria in the plains of India. Cytogenetic analysis of polytene chromosomes revealed that *A. culicifacies* is a complex of five sibling species designated as A, B, C, D, and E. Each sibling species has its own distribution pattern which is influenced by the environment. Sibling species differ in their larval and adult biology, response to insecticides, and capacity to transmit the disease. This knowledge has been applied in malariogenic stratification and the selection of cost-effective strategy in fighting malaria in India.

1. Introduction

Malaria is endemic in India. An estimated one billion population is at risk of malaria. Currently reported malaria incidence has stabilized at approximately 2 million parasite positive cases annually and deaths approximately 500–800. World Health Organization (WHO) estimates 15 million malaria cases and 20,000 deaths. Several independent studies and in-depth evaluations have brought out the fact that malaria incidence in the country is grossly underestimated (NMEP, 1985; Sharma, 1998). Malaria is unevenly distributed in time and space. Bulk of malaria cases are found in flood plains of northern India and in the east and west coastal plains. The northeastern region (population 28.5 million), and the forest and forest fringes on the hill ranges of peninsular India occupied by minority ethnic groups (population 71 million) have remained hyperendemic. Large parts of the country are visited by malaria epidemics, and in some years epidemics cover the entire ecotype like the epidemic of malaria in western Rajasthan in 1994 (Sharma, 1995). Malaria situation is still

---

V.P. Sharma • Centre for Rural Development and Technology (CRDT), Indian Institute of Technology (IIT), Hauz Khas, New Delhi-110016, India.

Malaria: Genetic and Evolutionary Aspects, edited by Krishna R. Dronamraju and Paolo Arese, Springer, New York, 2006.

alarming in Thar desert, stated to be sitting on the tip of the malarial iceberg (Sharma, 2004). The country is witnessing major ecological changes under the 5-year plans. Developmental activities have profoundly affected the epidemiology of malaria. Malaria till the 1950s was a rural disease, but over the last five to six decades new malaria ecotypes have emerged, such as man-made malaria in vast land areas under exploitation for settlement, industrial development, and agriculture (Pattanayak, *et al.* 1994; Sharma, 2002). Concurrently many new problems have emerged in the control of malaria viz., widespread vector resistance so that spraying now produces diminishing returns; emergence of drug resistance in *P. falciparum* and more recently in *Plasmodium. vivax* and deaths due to malaria that had been completely eliminated started visiting the endemic areas throughout the country (Dua *et al.*, 1996; Sharma, 1996; 1999). Resurgent malaria was widespread disseminating drug resistant strains. The return of malaria entered all receptive areas at one time freed from the disease (Sharma and Mehrotra, 1982a,b; 1986). The country now faces formidable challenges in malaria control viz., high cost of malaria control year after year, environmental pollution, resurgence of drug-resistant malaria, enormous losses in agriculture, industry, and tourism, difficulties in the development of the hinterland, rising trend of man-made malaria, epidemics which have become annual feature, and malaria which is the primary cause of illhealth in the endemic belts. Vector control is absolutely essential to control malaria effectively. Malaria-endemic areas remain poverty-stricken: malaria causes anemia, adversely affect the health of children and pregnant women, produces low birth babies, affects cognitive development in children, and is the single most important cause of poverty (Sharma, 2003).

There are about 450 Anopheles species in the world, of these 70 are considered vectors of malaria. The Indian anopheline fauna comprises of 58 species. Six anopheline are major vectors of malaria viz., *Anopheles culicifacies, A. stephensi, A. fluviatilis, A. minimus, A. dirus,* and *A. sundaicus* (Nagpal and Sharma, 1994) *Anopheles culicifacies* is the major malaria vector, widely distributed in the plains of India generating an estimated 65% malaria cases annually (Sharma, 1998) Several attempts to colonize *A. culicifacies* had failed, and as a result no advances could be made in our understanding of the basic biology of *A. culicifacies* (Russell and Rao, 1942). For the first time it was colonized by Ainsley (Ainsley, 1976) in Pakistan and India at the Malaria Research Centre (MRC) by Ansari *et al.* (1977). Currently the insectary at MRC maintains *inter alia, A. culicifacies sensu lato* originating from various places; sibling species A, B, C, and D; DDT and Malathion susceptible and resistant lines; and phenotypic and biochemical markers. Successful colonization of *A. culicifacies* opened up new opportunities for basic and applied research on *A. culicifacies.* Thus one of the important disciplines identified for thrust at the MRC was vector genetic research to control malaria in the country. In about two decades genetic research unfolded malaria epidemiology in diverse terrain and provided evidence for rational and sustainable malaria control. Work done on *A. culicifacies* is presented further.

Malaria control in India is largely the story of the control of *A. culicifacies*. This vector is active in malaria transmission in Iran, Afghanistan, Pakistan, India, and Sri Lanka. It prefers dry and semihumid climate. It is a monsoon-associated species and therefore bulk of malaria transmission occurs during rainy season. *A. culicifacies* breeds in the sunlit ponds and rainwater collections, newly dugout pits, natural and man-made water bodies, banks and beds of canals, subcanals, peripheral channels, seepage, small water collections, such as hoof prints and wheel rut and borrow pits, etc. Although rainfall encourages the breeding of all malaria vectors but it has profound impact on the breeding of *A. culicifacies* (Singh and Sharma, 2002). Therefore India's malaria is said to be a rainfall phenomenon. *A. culicifacies* has been incriminated in the transmission of *P. vivax, P. falciparum, P. malariae,* and drug-resistant malaria. The main obstacle in malaria control is the problem of physiological resistance to insecticides. An estimated 70–80% National Malaria Eradication Programme (NMEP's) efforts and cost of malaria control goes in the control of only one malaria vector (i.e., *A. Culicifacies*). This vector maintains endemic malaria throughout rural India and its rational control has remained an enigma. *A. culicifacies* transmitted malaria is highly uneven, unstable, and brings periodical epidemics (Raghavendra *et al.*, 1997; Sharma, 2002).

2. The *Anopheles culicifacies* Complex

A. culicifacies has high polymorphism as displayed by the presence of five sibling species. Green and Miles (1980) identified two sibling species in *A. culicifacies* designated as species A and B. This identification was based on two paracentric inversions a and b, on the X-chromosome in natural populations, in the apparent absence of heterozygotes. This was the first evidence of biologically distinct species within the taxon *A. culicifacies* Giles. Subsequently, within Xab population of *A. culicifacies* in Gujarat and Madhya Pradesh Subbarao *et al.* (1983) found two fixed paracentric inversions of g^1 and h^1 in chromosome arm 2. While species B was fixed for g^1 inversion (Xab; $2g^1 + h^1$) the new population with h^1 inversion was designated as species C (Xab; $2 + g^1h^1$). Furthermore, in a few populations in northern India, i^1 inversion was found polymorphic in species A population but in few places in northern and central India, a deficiency of heterozygotes was found for

i^1 inversion and in southern India a total absence of inversion heterozygotes was observed between species A ($X + a + b$; $2 + g^1 + h^1$) and the population with $X + a + b$; $2\ i^1 + h^1$ arrangement in certain areas (Suguna *et al.*, 1983; Subbarao, 1988). These observations provided population cytogenetic evidence for yet another species, species D ($X + a + b$; $2\ i^1 + h^1$) in *A. culicifacies* complex (Vasantha *et al.*, 1991). Figure 7.1 gives the cytomap of the four sibling species as worked out by the genetic research team at the MRC.

Surveys were undertaken throughout the country to map the distribution of *A. culicifacies* sibling species. It was revealed that species B is prevalent throughout India. In north India species A and species B are sympatric with predominance of species A (Subbarao *et al.*, 1987). In central India species A, B, C, and D occur in different proportions. In west India species B and C are sympatric with the predominance of species C. In southern India species A and B are sympatric with predominance of species B (Subbarao, 1988). For example, in Tamil Nadu 65–75% malaria cases are reported from 10 towns; and rural malaria is

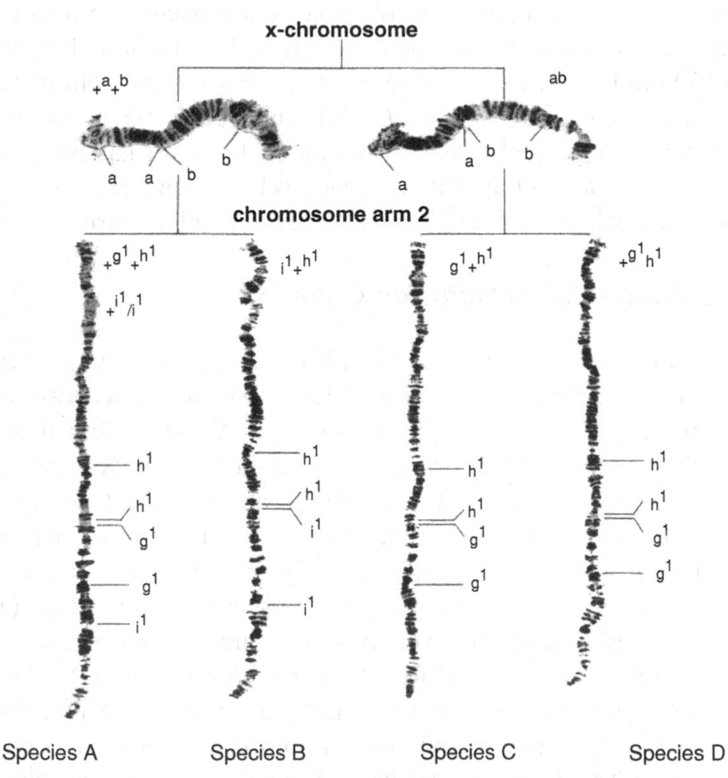

Figure 7.1. Paracentric inversions in polytene chromosomes of *A. culicifacies* sibling species A, B, C, and D. Reproduced with permission from Malaria Research Centre, New Delhi.

largely limited to areas with the prevalence of species E, or increase in species A populations due to irrigation projects. Results of the *A. culicifacies* sibling species distribution have been plotted in Figure 7.2.

The sibling species composition in any given area differs seasonally and with geographical location (e.g., species A is predominant in north India except in September–October) when in certain areas proportion of species B may exceed the proportion of species A (Subbarao *et al.*, 1987). *A. culicifacies sensu lato* is predominantly zoophagic species (Afridi, *et al.*, 1939; Bhatia and Krishnan, 1957). Studies on the host feeding pattern of *A. culicifacies s.l.* in UP and Bihar in 1985 and 1986 indicated relatively higher degree of anthropophagy for species A (e.g., in Ghaziabad, India) human blood index (HBI) of species A was 0.01 in May and this ratio increased to 0.08 in August and declined to 0.03–0.04 in October to November. In contrast HBI in species B remained zero in all months except in September and October when it was 0.02 (Joshi *et al.*, 1988). Anthropophagy provides an indication of

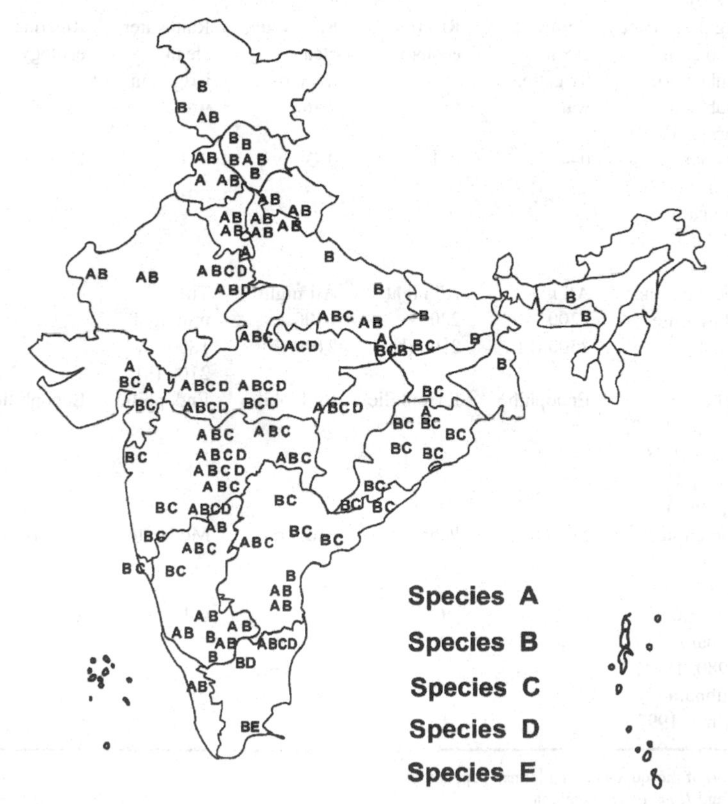

Figure 7.2. Distribution of *A. culicifacies* sibling species A, B, C, D, and E in India. Reproduced with permission from Malaria Research Centre, New Delhi.

the role of vectors species in malaria transmission (e.g., in *An. fluviatilis*), of the three sibling species designated as species S, T, and U; species S is the vector with high anthropophagy whereas species T and U are nonvectors/poor vectors and feed almost exclusively on animals (Subbarao *et al.*, 1994; Nanda *et al.*, 2000). Table 7.1 gives the biological characteristics of *A. culicifacies* as investigated over a period of 20 years (MRC, 2002). In *A. culicifacies,* each sibling species has a niche to

Table 7.1. Biological Variations Among Members of *A. culicifacies* Complex

Characteristics	Species A	Species B	Species C	Species D	Species E
Inversion Genotype (Green and Miles, 1980; Subbarao *et al.*, 1983; Vasantha *et al.*, 1991; Kar *et al.*, 1999).	X + a + b $2 + g^1 + h^1$ $+i^1/i^1$	Xab $2g^1 + h^1$	Xab $2 + g^1h^1$	X + a + b $2i^1 + h^1$	Xab $2g^1 + h^1$ Submetacentric Y-chromosome
Breeding preference (Subbarao *et al.*, 1987; Subbarao, 1988; Subbarao and Sharma, 1997)	Rainwater, clean irrigation water	Riverine ecology	Rainwater, clean irrigation water	Rainwater, clean irrigation water	Riverine ecology[c]
Anthropophilic[a] index (%) (Jambulingam *et al.*, 1984; Joshi *et al.*, 1988)	0–4	0–1	0–3	0–1	Up to 80[c]
Biting activity and peak biting time (Sathyanarayan, 1996)	All night 2200–2300 (h)	All night 2200–2300 (h)	All night 1800–2100 (h)	Till midnight 1800–2100 (h)	-
Resting behavior (Afridi *et al.*, 1939; Joshi *et al.*, 1988; Subbarao and Sharma, 1997)	Endophilic	Endophilic	Endophilic	Endophilic	Endophilic
Vector potential,[b] Subbarao and Sharma, 1997	Moderate	Poor	Moderate	Moderate	Very High
Sporozoite rate (%), Subbarao *et al.*, 1980, 1988, 1992; Subbarao and Sharma, 1997	0.51	0.04	0.3	0.4	20[c]

[a]Proportion of mosquitoes biting human population.
[b]*P. vivax* and *P. falciparum* malaria.
[c]In Rameshwaram island.
Source: Malaria Research Centre, New Delhi.

fill; they differ in biology, response to insecticides, biting rhythm, and capacity to transmit malaria.

Initially when the work on sibling species identification started, the only method to incriminate mosquitoes was the dissection of mosquitoes to demonstrate the oocyst/sporozoites infection (Subbarao *et al.*, 1980). This method was time consuming, cumbersome, and at times unreliable; required individual dissections and species identification of the parasite which was not possible. We trained the scientists of MRC in the newly developed technique of two-site immunoradiometric assay (IRMA) to detect infected mosquitoes as part of the Indo–US bilateral collaboration. In this technique species-specific monoclonal antibodies were used against the circumsporozoite (CS) proteins of *P. vivax* and *P. falciparum* (Zavala *et al.*, 1982). Mosquitoes were subjected to cytological identification of sibling species, and the identified specimens were processed for the detection of CS antigen by IRMA/Elisa. These studies brought out that *A. culicifacies* species A mosquitoes were gland positive for *P. vivax* and *P. falciparum* in the villages in UP, Haryana, and Delhi. Species B specimens were not found gut or gland positive in villages around Delhi and in other places in the western India (Subbarao *et al.*, 1980; Subbarao *et al.*, 1987). This study was extended to central India in the state of Madhya Pradesh. This region has perennial malaria transmission despite DDT and HCH spraying. Surveys in Kundam block, Jabalpur district and Bijadhandi block, Mandla district in 1988 revealed that all four sibling species were present. Species C had the highest proportion (65–90%) followed by species D (6–31%), and both were found with sporozoites by IRMA; while species A and B were found in very low numbers (Subbarao *et al.*, 1992). Further evidence of relative susceptibility of *A. culicifacies* species A, B, and C was provided by laboratory feeding of mosquitoes on *P. vivax*-infected donors through artificial membrane and comparing it with *A. stephensi*. In this experiment species A had the highest percentage of mosquitoes with oocysts and sporozoites followed by species C, and species B was least susceptible. Among all fed mosquitoes >5% species B was infected compared with 26% species C, and 63% species A and 77% *A. stephensi* (Adak *et al.*, 1999). Susceptibility of *A. culicifacies* species A, B, and C were tested on rodents infected with *Plasmodium yoelii yoelii* and *Plasmodium vinckei petteri*. Species A and species C specimens had the highest number of mosquitoes with sporozoites (i.e., 23.1% and 22.7% respectively) whereas species B specimens had B 5.3%, the lowest. Positivity for sporozoites in A. stephensi held as control ranged from 53.1 to 57.9% (Kaur et al., 2000).

Changes in local ecology (e.g., canal irrigation) favors species A and spraying in agriculture favors species C; and both are malaria vectors. Further species A and C thrive better in premonsoon conditions and prefers to breed in nonriverine and irrigated terrain, whereas species B thrives better in postmonsoon conditions and prefers to breed in riverine terrain (Subbarao and Sharma, 1997). In areas with sympatric populations of species A and B, the two sibling species do not have any specific preference for resting habits (Subbarao *et al.*, 1987). Behavior of mosquitoes is important for planning malaria control. For example in areas under the influence of *A. culicifacies s.l.* biting goes on all night, but species C and D bite early (peak activity time 1800–2100 h); whereas species A and B biting activity starts late (peak activity time 2200–2300 h) (Sathyanarayan, 1996). This behavior influences the success of insecticide-treated mosquito net programme (Yadav *et al.*, 2001).

2.1. Paradox Resolved

During the malaria eradication phase of the NMEP, districts in western UP were hyperendemic and required DDT spraying (two rounds @1 g/sq m) to control malaria. In contrast, eastern UP and north Bihar districts were hypoendemic and required one round of DDT spraying (@1 g/sq m) to control malaria. Ramachandra R.T. (1984) stated that one of the unresolved paradoxes in the epidemiology of malaria is related to the differential vectorial potential of *A. culicifacies* in regions of apparently similar physiography. This paradox of the changing profile of malaria was difficult to understand particularly when *A. culicifacies s.l.* was the only vector of any importance in malaria transmission in the entire belt from west to east in north India (NMEP, 1960). First indication of the possible variations in the vector potential came with the discovery of species A and B in *A. culicifacies s.l.* Incrimination studies of species A and B revealed that species A was the vector and species B was nonvector (Subbarao *et al.*, 1988). Further surveys on the prevalence of sibling species in UP and Bihar revealed that species A was the predominant species in western districts of UP and Uttranchal. The proportion of species A declined sharply in the eastern districts and north Bihar. As per the vector incrimination work, the incidence of malaria should coincide with the prevalence of species A. This hypothesis was tested in UP and Bihar in early 1986. When this picture was examined in the background of malaria incidence reported by the NMEP, it was revealed that malaria transmission occurs only in areas with predominance of species A. Figure 7.3 gives the results of malaria incidence and sibling species composition in study districts in UP and Bihar. For example, in Bulandshahr, Ghaziabad, Shahjahanpur, and Nainital malaria transmission was perennial with periodical epidemic situations causing high mor-

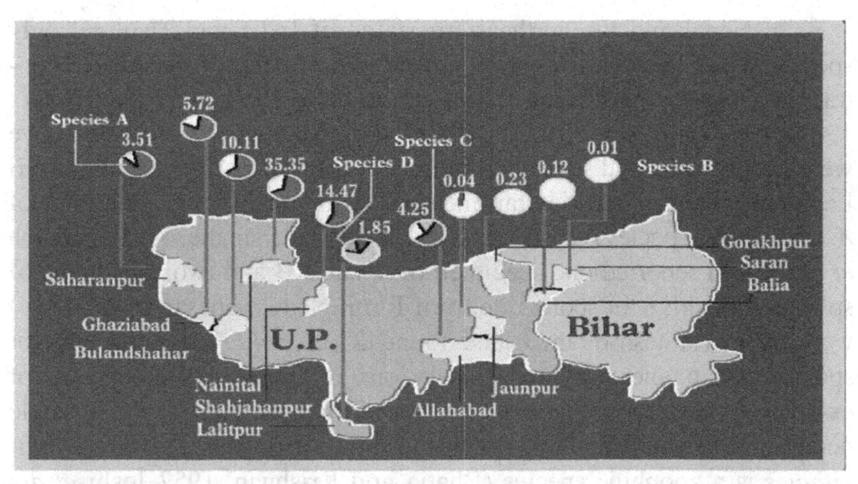

Figure 7.3. Distribution of *A. culicifacies* in UP and Bihar with circles showing the proportion of sibling species in each place surveyed. Arabic figures give the annual parasite incidence (API) in different districts. See the high API in districts with species A, C, and D and low to nil API in districts with exclusive population of species B. Reproduced with permission from Malaria Research Centre, New Delhi.

bidity and deaths due to malaria. Specimens collected from the field revealed that the proportion of species A exceeded 50% and since this species has high vectorial potential, malaria Annual Parasite Incidence (API) varied from 5.72 to 35.35 in the western UP districts. In contrast to the eastern districts and north Bihar, malaria prevalence was very low (proportion of species B was 90–100%). In Jaunpur, malaria incidence was low except in May when slide positivity rate was 10.3 (species A was 2–8%). In Balia, malaria cases were few (API 0.12) and in Saran, no malaria case was reported. It is noteworthy to mention that *P. vivax* and *P. falciparum* cases are being brought in to eastern UP and Bihar villages by the labor who returns from malaria-endemic areas in Gujarat, Rajasthan, Punjab, and Maharashtra in April and May, and this labor returns for work in August. The parasite is therefore present in the population but there is no indigenous transmission, clearly indicating the lack of vectorial capacity in allopatric populations of species B. It may be noted that in eastern districts of UP and north Bihar, spraying has been withdrawn since 1964, although *A. culicifacies* is present in abundance (Subbarao *et al.*, 1988). Malaria cases have remained very low and no malaria outbreaks have been reported from these areas, further confirming that areas with allopatric population of species B enjoy freedom from malaria. Thus genetic research has resolved this paradox and enhanced our understanding of malaria transmission dynamics in the country.

In Nepal and the northeastern states of India and Thailand only species B was found, which is not considered a vector of malaria. In contrast, *A. culicifacies* species A was incriminated in Arabia (Akoh *et al.*, 1984), Iran (Zaim *et al.*, 1993) and Pakistan (Mahmood *et al.*, 1984). We were intrigued to find several reports from Rameshwaram Island and Sri Lanka of intense malaria transmission with exclusive populations of *A. culicifacies* species B (Sabesan *et al.*, 1984; Amersinghe, 1991). Suguna *et al.* (1983) reported one specimen of species B and two specimens of species A positive for sporozoite from Tamil Nadu. Sabesan *et al.* (1984) working in Rameshwaram Island found 21 *A. culicifacies* positive for sporozoites in one year, and populations of this island comprised of exclusively species B. The evidence started building up for the possible occurrence of another sibling species in the *A. culicifacies* taxon. *A. culicifacies* is a zoophilic species (Bhatia and Krishnan, 1957 Joshi *et al.*, 1988; Nanda *et al.*, 2000), but feeding experiments in Rameshwaram island revealed that *A. culicifacies* simultaneously exposed on man and cattle preferred feeding on man (Jambulingam *et al.*, 1984), indicating a sharp deviation in species B feeding preference from populations in north India. This led to further studies on the possible occurrence of another member of sibling species, and indeed sibling species E was discovered within the species B. The Y-chromosome of species E is submetacentric whereas that of species B is acrocentric (Kar *et al.* 1999; Surendran *et al.*, 2000). Interestingly, differentiation of *A. culicifacies* based on biochemical methods and DNA probes have limitations and, the male mitotic karyotype is unreliable (Adak, 1997). Therefore the diagnostic inversions on polytene chromosomes are still the most accurate methods of sibling species identification in *A. culicifacies*. This evidence is lacking in species E and work is in progress to resolve this in species E.

3. Application in Malaria Control

Rural malaria control in India is based on the residual spraying of insecticides like DDT, HCH, Malathion, and synthetic pyrethroid insecticides. DDT is the principal insecticide used in malaria control. In 2002, 40 million populations were under DDT spraying, although Indian DDT does not meet the WHO specifications, and the dosage sprayed is 1 g instead of 2 g/sq m as recommended by the WHO. In vast areas of the country DDT is not effective in malaria control (Sharma, 2003). As per the NMEP policy DDT should be the first insecticide of choice and sprayed as long as it offers an epidemiological impact, irrespective of the vector susceptibility status to it. When DDT resistance is confirmed by WHO test procedures, and it is also confirmed that DDT spraying had no epidemiological impact on malaria incidence, in spite of good quality and coverage, a change of insecticide is proposed. This criterion is used for all other replacement insecticides. Hexachlorocyclohexane has been banned. Malathion should be sprayed in areas with DDT resistant vector populations. Malathion spraying has poor house coverage because of its pungent odor. Synthetic pyrethroids (SP) are being used increasingly in malaria vector control. However, there are reports of resistance to SP insecticides, and in the coming decade, widespread resistance to SP is likely to result in control failures (Singh *et al.*, 2002; Mittal *et al.*, 2002). Insecticide resistance would then become the most important determinant of malaria in rural India (Sharma, 2002). Our work in vector genetics has provided scientific basis for the change of insecticides and their selective application to achieve cost-effective malaria control. A brief description of the inputs provided to the NMEP is given below.

3.1. Stratification

Malaria control in the country is based on broad stratification of the country in respect of vector distribution, susceptibility to insecticides, impact on malaria transmission, and availability of resources for its control. Genetic research has provided additional tool for stratification to optimize interventions. Sibling species based stratification would result in reduced burden of insecticides in the environment and produce cost effective and sustainable malaria control. Knowledge of sibling species should be applied at the national level in planning supplies for interventions. At the state level, districts, or Primary Health Centres (PHCs) it could be stratified broadly for the type of interventions most appropriate for each area. A further stratification would be required at the district and PHC level to target most location-specific interventions. For example, in north India with predominantly species A, spraying DDT would produce good epidemiological impact on malaria. In contrast. in the eastern Uttar Pradesh and north Bihar spraying insecticide would be a complete waste. These areas lack indigenous transmission and malaria control should rely on early case detection and prompt treatment (EDPT). In contrast, in the western and central districts with predominance of species A, Malathion could be sprayed as a short-term strategy switching over to environmental interventions. In districts prone to malaria epidemics, integrated malaria control methods should be applied as a cost-effective strategy. In south Indian districts with low malaria prevalence due to predominance of species B, bioenvironmental interventions would be the most appropriate strategy (Sharma,1987; 1988; Dua *et al.*, 1988; Sharma, 1991; Subbarao and Sharma, 1994; Sharma, 1998). Spraying

Hexachlorocyclohexane (HCH, now banned) is not recommended in any part of the country, as *A. culicifacies* develops resistance rather quickly. However, introduction of Lindane (gamma isomer of HCH) is currently under debate. It has been introduced in agriculture. If Lindane is sprayed to control *A. culicifacies*, the vector populations would develop resistance rather quickly as they are already resistant to HCH.

At the district level stratification based on sibling species prevalence would be very useful in targeted selective interventions (e.g., the malariogenic stratification of Allahabad rural, in India). The Ganga and the Yamuna rivers divide this region into three well-defined zones known as Gangapar, Yamunapar, and Doaba. In Gangapar and Doaba, malaria transmission is always low to nil. This is because of the high prevalence of species B (65.6% in Gangapar and 88% in Doaba). In Yamunapar with high malaria incidence, the overall composition of species B was 57.7%, similar to Gangapar and Doaba. Further stratification of Yamunapar revealed that Shankargarh block was responsible for bulk of malaria cases in the Yamunapar area. The sibling species composition in Shankargarh block comprised of species A (64%), species B (25%), and species C (11%), and together the two species A and C accounted for 75% *A. culicifacies*. Prevalence of malaria had a direct relationship with the presence of vector sibling species A and C. Thus the revised strategy to control was to spray Shankargarh block, instead of the entire Yamunapar zone. This has saved insecticides and operational cost of malaria control and prevented pollution (Tewari *et al.*, 1994).

3.2. Insecticide Resistance

Figure 7.4 gives the present status of insecticide resistance in malaria vector *A. culicifacies. An. culicifacies sensu lato* has become triple resistant (DDT, HCH, and Malathion) with problem of chloroquine resistance in 26 million populations. Additionally, 78.9 million population live in triple vector resistant areas in Andhra Pradesh, Bihar, Gujarat, Madhya Pradesh, Maharashtra, Orissa, and Rajasthan. Priority in spraying is given to areas under epidemics or impending epidemics, natural disasters, project areas of high economic importance like irrigation, river valley projects, industrial units, and areas with increasing number of deaths due to malaria. Vector control in these areas is one of the most difficult challenges in malaria control. Table 7.2 gives the impact of DDT, HCH, Malathion, and SP on the development of resistance in *A. culicifacies* sibling species in the field.

National Malaria Eradication Programme (NMEP) was reporting that *A. culicifacies s.l.* is resistant to DDT, yet its spraying produces epi-

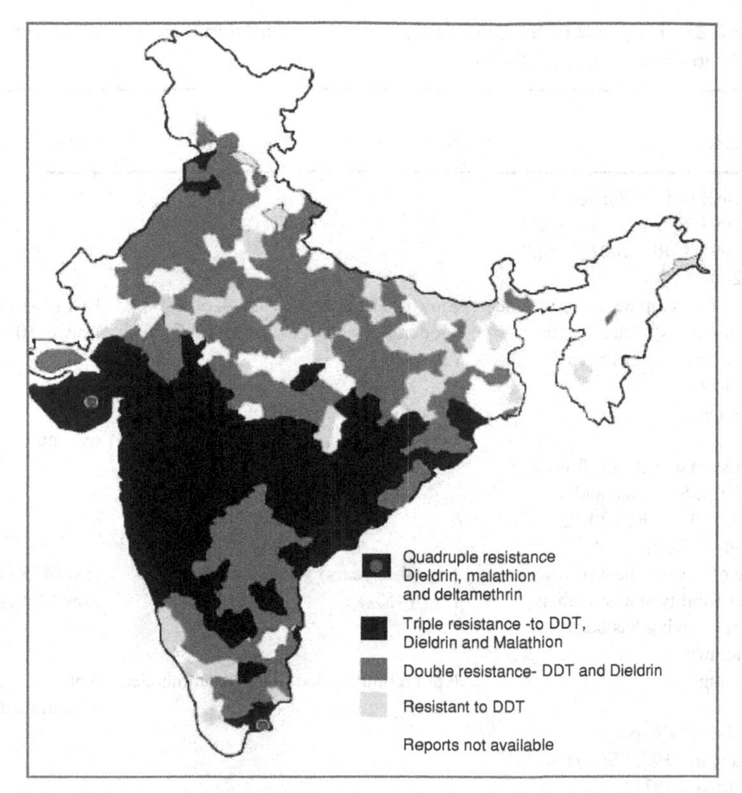

Figure 7.4. Current status of insecticide resistance in *A. culicifacies* in India. *An. culicifacies* is resistant to DDT in the entire country, to DDT and HCH in almost all parts of the country, and to DDT, HCH, and Malathion in Gujarat, Maharashtra, parts of Madhya Pradesh, Rajasthan, and in few other states; and to Deltamethrin in Gujarat, Maharashtra, Rameshwaram island, and Tamil Nadu. Reproduced with permission from Malaria Research Centre, New Delhi.

demiological impact on malaria. Sharma *et al.* (1982, 1986) demonstrated by spraying DDT in DDT-resistant *A. culicifacies* populations and reported a similar observation. This phenomenon was explained by testing susceptibility of species A and species B to DDT (1 h exposure on 4% DDT papers). Results revealed that species A was more susceptible to DDT than species B (Afridi *et al.*, 1939; Subbarao and Sharma, 1997). Species A has high capacity to transmit malaria, but development of resistance to DDT and Malathion is slow (i.e., emergence of resistance to DDT in areas with species A took >9–10 years of DDT spraying @ 1 g/sq m, two rounds under the NMEP). In case of Malathion, emergence of resistance took 9–10 years of spraying @ 2 g/sq m, three rounds in Sonepat, Haryana, India (Subbarao, 1988; Subbarao and Sharma, 1994; 1997). In contrast DDT spraying under the NMEP resulted in resistance

Table 7.2. Response of Sibling Species to Insecticides and Recommended[a] Spray Strategy to Control *A. culicifacies*

Insecticides	Species A vector	Species B[b] nonvector	Species C vector
DDT (Subbarao, 1988; Tewari *et al.*, 1994; Subbarao and Sharma, 1994; 1997; MRC, 2002; Mittal *et al.*, 2002)			
• Rate of development of resistance	Slow (9–10 years)	Fast (4–5 years)	Fast (4–5 years)
• Susceptibility status in areas where spraying has been withdrawn	Moderate (30–60%)	Low (<10%)	Low (<10%)
• Spraying	Recommended	Not required	Not recommended
HCH[c] (Subbarao, 1988; Tewari *et al.*, 1994; Subbarao and Sharma, 1997; MRC, 2002; Mittal *et al.*, 2002)			
• Rate of development of resistance	Fast (4–5 years)	Fast (4–5 years)	Fast (4–5 years)
• Susceptibility status in areas where spraying has been withdrawn	Low (<5%)	Low (<5%)	Low (<5%)
• Spraying	Not recommended	Not recommended	Not recommended
Malathion[d] (Rajgopal, 1977; Sharma *et al.*, 1986; Subbarao and Sharma, 1997)			
• Rate of development of resistance	Slow (9–10 years)	Intermediate	Fast (4–5 years)
• Susceptibility status in areas where spraying has been withdrawn	No evidence	(6–7 years) Not required	Recommended, if susceptible to Malathion
• Spraying			
Carbamates/Synthetic Pyrethroids[d] (Subbarao and Sharma, 1994; 1997; Singh *et al.*, 2002; Mittal *et al.*, 2002)			
• Susceptibility	High	High	High
• Development of resistance due to usage in agriculture	No evidence	No evidence	
• Spraying	Not necessary	Not required	Recommended if resistant to Malathion

[a]The strategy proposed is for all sibling species. As species D has very limited distribution, and is always in sympatricity with other species in low proportions, no specific strategy is being recommended. Depending on the predominance of species A or C, specific strategy for that species may be followed.
[b]In a few areas where species E is implicated as a vector of malaria such as in the Rameshwaram island. The existing spray strategy to be continued and follow the recommendation as given under[d].
[c]Resistance may precipitate after a few rounds of spray as the populations are not fully susceptible. Use Malathion as focal spray.
[d](i) Spray for 4 to 6 rounds, usually incidence comes down, and withdraw the insecticide, (ii) In the following years use a new insecticide or preferably an insecticide that is chemically unrelated to the one in use in focal spraying, and (iii) Resume the spraying whenever there is an increase in the incidence.
Source: Malaria Research Centre, New Delhi

in species C in Surat (Gujarat). In Surat, spraying of Malathion in 1969 onwards to interrupt malaria transmission resulted in Malathion resistance in *A. culicifacies* in 4–5 years (Rajgopal, 1977). In Sonepat, Haryana, malaria transmission was intense despite of DDT and HCH spraying to control malaria. Spraying three rounds of Malathion continued for 7–8 years although there was no evidence of malaria transmission, although its spraying was required only after the resumption of transmission. Unfortunately, spraying continued for over a period of 8–9 years, and eventually species A developed resistance to Malathion (Raghavendra *et al.* 1991; 1992; Sathyanarayan, 1996).

In Andhra Pradesh, Warangal, Khammam, and Mahabubnagar districts farmers were spraying organophosphate insecticides to protect cash crops such as the chillies, cotton, spices, etc. This practice induced resistance in the vectors that breed in agricultural fields (e.g., *A. culicifacies*). Even before the Malathion spraying started, *A. culicifacies* was found resistant to Malathion (90–100% adult survival at 1 h exposure on 5% Malathion papers). Since many of the insecticides used in agriculture have carboxyl esterase mediated resistance mechanism, *A. culicifacies* eventually became resistant to Malathion. In Andhra Pradesh and Gujarat, *A. culicifacies* sibling species B (nonvector species) and species C (malaria vector) are prevalent. A particularly disturbing feature of this phenomenon was that *A. culicifacies* species C was in very low proportions (2–15%) before the onset of Malathion resistance and as the resistance progressed species C populations increased to 60–70% in Andhra Pradesh and 60–90% in Gujarat, thus enhancing malaria transmission (Raghavendra *et al.* 1991; 1992). Resistance in species B builds up rapidly to DDT and synthetic pyrethroids but takes little longer in case of Malathion. Malaria control in species C areas by spraying DDT and Malathion is unlikely to be productive in the long run, and therefore NMEP should rely on alternative strategies, such as the bioenvironmental methods and insecticide-treated mosquito nets. It may be noted that developmental changes result in the mosquito fauna changes and composition of the sibling species. Irrigation encourages the breeding of species A. The DDT spraying suppresses the sibling species A and increases the population of species B due to rapid development of resistance. Malathion spraying changes the composition in favor of species C and resistance to Malathion develops rather quickly in vectors due to spraying to protect crops. Populations of species D are very limited; and very little is known about species E. Therefore, it is important to study local malaria transmission and the role of each sibling species for cost-effective malaria control. Such information is critical in the planning of malaria control.

3.3. Bioenvironmental Malaria Control

Bioenvironmental malaria control strategy is principally based on the reduction or elimination of mosquito breeding habitats by environmental management and biological control methods. In the area selected for malaria control, geographical reconnaissance provides detailed information on the possible mosquito-breeding sites. Appropriate interventions are applied on each site to ensure that mosquito breeding would not occur in that habitat. The interventions generally involve drainage of water bodies that are not useful either for agriculture, wild life, or human use. Small water bodies are eliminated by earthwork particularly low-lying areas, borrowpits, ditches, excavations, etc. Expanded polystyrene beads are applied in wells and tanks. In marshy areas, poplar and eucalyptus plantations can dry the surface water. This is particularly useful for wasteland. Water bodies essential for the communities and wildlife can be treated with fishes. *Gambusia affinis* popularly known as Gambusia or mosquito fish and *Poecilia reticulata* known as Guppy fishes can be released in such water bodies after ensuring that there are no predatory fishes and margins of the ponds have been cleaned of weeds. In all these activities, communities are actively involved so that the beneficiaries share part of the intervention efforts. For this intensive health education campaigns are organized. The first community-based bioenvironmental malaria control project was launched in 1984 in Nadiad Taluka, district Kheda in Gujarat. In these villages, an epidemic of malaria had killed 35 persons in Bamrauli village alone. Malaria transmission was intense even though villages had been sprayed with DDT and Malathion. Indoor residual spraying (IRS) had failed in malaria control. This was not an isolated report but such failures had become commonplace. At that time there was no alternative strategy to control malaria in the country. Results in the first year of interventions were spectacular and malaria incidence was reduced by about 94%, and in subsequent years malaria transmission was interrupted (Sharma and Sharma 1989). An international group of experts drawn from WHO, and other national and international organizations evaluated the project. The strategy was found highly cost effective and sustainable, and required more research for its extension to other malaria ecotypes (Anonymous, 1987). The government of India provided additional funds to study the feasibility of malaria control in situations with problems related to vectors, parasite(s), human ecology, and cost effectiveness. Twelve locations were identified to demonstrate the feasibility of bioenvironmental methods in malaria control. These field studies led to the successful demonstration of malaria control in rural, urban, industrial, tribal, and coastal areas of the country (Sharma, 1987; 1988; Dua *et al.*, 1991; Singh and Sharma, 2000). A major challenge still remained, of the demonstration of this malaria control strategy through the existing primary health care system.

This challenge was taken up in Karnataka state jointly with the state health department. Kolar district was selected for demonstration. In this region of Karnataka, sericulture is the main cottage industry, and farming community does not allow any indoor residual spraying of insecticide for fear of toxicity to the silk worms. As a result, malaria was rampant and malaria API was as high as 700–800 in some villages. Deaths due to malaria were common. A team of MRC was deputed to work jointly with the state health department in this challenging terrain. Initially, all staff of the health department was trained by organizing workshops, and hatcheries were established for the supply of Guppy fishes. Geographical reconnaissance showed that *A. culicifacies* largely breeds either the streams or irrigation wells. While managing of wells was easy, but control of mosquito breeding in streams appeared almost impossible. These streams go around most of the villages and measure 200–500 m in width with huge boulders embedded in a rocky bed. Survival of Guppy and Gambusia in these streams with high water speed appeared impossible. At this point we took the help of our work on the sibling species complex. Analysis of field-collected specimens from the area revealed the presence of *A. culicifacies* species A (vector) and species B (nonvector), and *A. fluviatilis* species T (nonvector). Irrigation wells were supporting the breeding of species A whereas streams had species B as the most dominant species. We therefore decided to focus our work on the control of mosquito breeding in wells. Initially we validated this finding by selecting three sets of villages in Kamasanudram PHC villages:

(i) Puram village (population 396) with 22 wells and no steams;
(ii) Bodapatti village (population 396) with four wells and a perennial stream, and
(iii) Banganatham village (population 130) situated on a perennial stream and had no wells.

Malaria incidence before the introduction of Guppy fishes from 1994 to 1996 was: Puram village API was 50.3, 52.8, and 120.6 (cases 20, 21, and 48), respectively. For the same year malaria API in Bodapatti village was 0.2, 0.2, and 0.7 (cases 1, 1, and 3) and in Banganatham village was 1.6, 1.6, and 0.8 (cases 2, 2, and 1). The prevailing malaria situation indicated that malaria incidence had a direct relationship with the presence of wells that were the source of species A. In 1996, Guppy fishes were introduced in wells and streams.

In 1997, malaria API was reduced to 27.6 (11 cases) in Puram, 1.6 (2 cases); there was no case in Bodapatti, and API 1.6 (2 cases) in Banganatham. In subsequent years from 1998 to 2003, no case of malaria was reported from Puram, Bodapatti, and Banganatham. This fieldwork encouraged us to expand the programme to cover more areas in Karnataka (Ghosh *et al.*, 2004). At present 400 villages are under the bioenvironmental interventions and there is no evidence of indigenous transmission from these areas. The mosquito populations have also declined making the life of the rural community comfortable in the night. Genetic research provided a scientific basis for launching the bioenvironmental malaria control in Karnataka. Currently, bioenvironmental malaria control program is under implementation through the primary health care system in >150 million population, and the programme is under expansion to cover more areas and diseases (e.g., to Japanese encephalitis and Dengue fever etc). Malaria control is in the maintenance phase or under intensive expansion to reach all villages in Kolar and Hassan Districts, Karnataka; the entire state of Maharashtra; 100 predominantly tribal districts under the Enhanced Malaria Control Project of the NMEP; Madhya Pradesh; Gujarat; Goa; urban areas in Chennai, Ahmedabad, Mangalore; and a few other areas. Impact of interventions has reduced malaria and mosquito nuisance and the workers are filled with enthusiasm, not seen in the spray squads. There have been no outbreaks of malaria as against the spraying of DDT, Malathion, or SP. In these areas, malaria returns every year with high potential requiring preventive spraying. If, for reasons beyond the control of NMEP, these areas are not sprayed, malaria cases starts building up culminating in epidemics. In some areas even in the presence of spraying malaria outbreaks are being witnessed. As against the indoor residual spraying of insecticides, the bioenvironmental malaria control strategy has emerged as the most appropriate, cost-effective, and sustainable method of malaria control in the country (Sharma,1987; 1988; Dua *et al.*, 1988; Sharma, 1991; Subbarao and Sharma, 1994; Sharma, 1998).

4. Acknowledgments

The included in this chapter was carried out by the scientific and technical staff of the Malaria Research Centre, Delhi and I shall ever remain grateful for their dedication, honesty and commitment. I wish also to thank Dr. T. Adak and Dr Nutan Nanda who had kindly reviewed the manuscript and made important suggestions. Thanks are also due to Shri Tarun Mehrotra for the art work.

References

Adak, T., Kaur, S., Wattal, S., Nanda, N., and Sharma, V.P. (1997). Y-chromosome polymorphism in species B and C of *Anopheles culicifacies* complex. *J. Am. Mosq. Contr. Assoc.*, **13**(4), 379–383.

Adak, T., Kaur, S., and Singh, O.P. (1999). Comparative susceptibility of different members of *Anopheles culicifacies* complex to *Plasmodium vivax. Trans. R. Soc. Trop. Med. Hyg.*, **93,** 573–577.

Afridi, M.K., Singh, J., and Singh, S.H. (1939). Food preferences of *Anopheles* mosquitoes in the Delhi urban area. *J. Mal. Inst. India*, **2,** 219–228.

Ainsley, R.W. (1976). Laboratory colonization of the malaria vector *Anopheles culicifacies. Mos. News,* **36,** 256–258.

Akoh, J.L., Beidas, M.F., and White, G.B. (1984). Cytoplasmic evidence for the malaria vector species A of the *Anopheles culicifacies* complex being endemic in Arabia. *Trans. R. Soc. Trop. Med. Hyg.*, **78,** 698.

Amersinghe, F.P., Amersinghe, P.H., Peiris, J.S.M., and Wirtz R.A. (1991). *Anopheles* ecology and malaria infection during the irrigation development of an area of the Mahaweli project Sri Lanka. *Am. J. Trop. Med. Hyg.*, **45,** 226–235.

Anonymous, (1987). In-depth evaluation of the community-based integrated vector control of malaria project in Kheda (Gujarat), Malaria Research Centre, New Delhi.

Ansari, M.A., Mani, T.R., and Sharma, V.P. (1977). A preliminary note on the colonization of *Anopheles culicifacies* Giles. *J. Com. Dis.*, **9**(3), 206–207.

Bhatia, M.L. and Krishnan, K.S. (1957). *Anopheles culicifacies* Giles 1901, in: *Vectors of Malaria in India*. The National Society for Malaria and Other Communicable Diseases, Delhi. pp. 108–140.

Dua, V.K., Kar, P.K., and Sharma, V.P. (1996). Chloroquine resistant *Plasmodium vivax* malaria in India. *Trop. Med. Int. Hlth.*, **1,** 816–819.

Dua, V.K., Sharma, S.K., and Sharma, V.P. (1988). Bioenvironmental control of malaria in an industrial complex at Hardwar (UP), India. *J. Am. Mosq. Contr. Assoc.*, **4,** 426–430.

Dua, V.K., Sharma, S.K., and Sharma, V.P. (1991). Bioenvironmental control of malaria at the Indian Drugs and Pharmaceuticals Ltd. Rishikesh (UP). *Indian J. Malariol.*, **28,** 227–236.

Ghosh, S.K., Tiwari, S.N., Sathyanarayan, T.S., Sampath, T.R.R., Sharma, V.P., Nanda, N., Joshi, H. Adak, T., and Subbarao, S.K. (2004). Larvivorous fish in well target the malaria vector sibling species of the *Anopheles culicifacies* complex in villages in Karnataka, India. *Trans. R. Soc. Trop. Med. Hyg.*, (in Press).

Ghosh, S.K et al. in Trans. R. Soc. Trop. Med. Hyg., 99(2):101–105.

Green, C.A. and Miles, S.J. (1980). Chromosomal evidence for sibling species of the malaria vector *Anopheles culicifacies* Giles. *J. Trop. Med. Hyg.*, **83,** 75–78.

Joshi, H., Vasantha, K., Subbarao, S.K., and Sharma, V.P. (1988). Host feeding patterns of *Anopheles culicifacies* species A and B. *J. Am. Mosq. Contr. Assoc.*, **4**(3), 248–251.

Jambulingam, P., Sabesan, S., Vijayan, V.A., Krishna Moorthy, K., Gunasakheran, K., Rajendran, G., Chandrahas, R.K., and Rajagopalan, P.K. (1984). Density and biting behavior of *Anopheles culicifacies* Giles in Rameshwaram island (Tamil Nadu). *Ind. J. Med. Res.*, **80,** 47–50.

Kar, I., Subbarao, S.K., Eapen, A., Ravindran, J., Sathyanarayan, T.S., Raghavendra, K., Nanda, N., and Sharma, V.P. (1999). Evidence for a new malaria vector species, species E, within the *Anopheles culicifacies* complex (Diptera: Culicidae). *J. Med. Entomol.*, **36**(5), 595–600.

Kaur, S., Singh, O.P. and Adak, T. (2000). Susceptibility of species A, B and C of *Anopheles culicifacies complex to Plasmodium yoelii yoelii* and *Plasmodium vinkei petteri infections. J. parasitiol* **86**: 1345–1348.

Mahmood, F., Sakai, R.K., and Akhtar, K. (1984). Vector incrimination studies and observations on species A and B of the taxon *Anopheles culicifacies* in Pakistan. *Trans. R. Soc. Trop. Med. Hyg.*, **78,** 607–616.

Mittal, P.K., Adak, T., Singh, O.P., Raghavendra, K., and Subbarao, S.K. (2002). Reduced susceptibility to Deltamethrin in *Anopheles culicifacies s.l.* in district Ramanathapuram in Tamil Nadu: Selection of pyrethroid resistant strain. *Curr. Sci.*, **82**(2), 185–188.

MRC. *A profile of Malaria Research Centre* (1977–2002). (2002). Malaria Research Centre (Indian Council of Medical Research), Delhi, India. pp. 1–234.
Nagpal, B.N. and Sharma, V.P. (1994). *Indian Anophelines*. Oxford and IBH, New Delhi.
Neeru, S. and Sharma, V.P. (2000). Malaria control in Madhya Pradesh, India. *Public Health*, **15,** 57–68.
NMEP. (1960). *Manual of the malaria eradication operation*. Directorate of NMEP, Government of India, Ministry of Health Publication, Delhi.
NMEP. (1985). *In-depth evaluation report of the modified plan of operation under National Malaria Eradication Programme of India*. Government of India Publication. pp. 1–56.
Nutan, N., Yadav, R.S., Subbarao, S.K., Hema J., and Sharma, V.P. (2000). Studies on *Anopheles fluviatilis* and *Anopheles culicifacies* sibling species in relation to malaria in forested and deforested riverine ecosystems in northern Orissa, India. *J. Am. Mosq. Contr. Assoc*., **16**(3), 199–205.
Pattanayak, S., Sharma, V.P., Kalra, N.L., Orlov, V.S., and Sharma, R.S. (1994). Malaria paradigms in India and control strategies. *Ind. J. Malariol*., **31,** 141–195.
Raghavendra, K., Subbarao, S.K. Vasantha, K., Pillai, M.K.K., and Sharma, V.P. (1992). Differential selection of Malathion resistance in *Anopheles culicifacies* species A and B (Diptera: Culicidae) in Haryana state, India. *J. Med. Entomol*., **29,** 183–187.
Raghavendra, K., Subbarao, S.K., and Sharma, V.P. (1997). An investigation into the recent malaria outbreak in district Gurgaon, Haryana, India. *Curr. Sci*., **73**(9), 766–770.
Raghavendra, K., Vasantha, K., Subbarao, S.K., Pillai, M.K.K., and Sharma, V.P. (1991). Resistance in *Anopheles culicifacies* sibling species B and C to Malathion in Andhra Pradesh and Gujarat state, India. *J. Am. Mosq. Contr. Assoc*., **7,** 255–259.
Rajgopal, R. (1977). Malathion resistance in *Anopheles culicifacies* in Gujarat. *Ind. J. Med. Res*., **66,** 27–28.
Ramachandra Rao, T. (1984). *The Anophelines of India*. Malaria Research Centre (ICMR) Publication, New Delhi.
Russell, P.F. and Ramachandra, Rao, T. (1942). On the swarming mating and oviposition behavior of *Anopheles culicifacies*. *Am. J. Trop. Med*., **22,** 417–427.
Sabesan, S., Jambulingam, P., Krishnamoorthy, K., Vijayan, V.A., Gunasekaran, K., and Rajendran, G. (1984). Natural infection and vectorial capacity of *Anopheles culicifacies* Giles in Rameshwaram Island. *Ind. J. Med. Res*., **80,** 43–46.
Sarbjit K., Singh, O.P., and Adak, T. (2000). Susceptibility of species A, B, and C of *Anopheles culicifacies* complex to *Plasmodium yoelii* and *Plasmodium vinckei petteri* infections. *J. Parasitology*, **86**(6), 1345–1348.
Sathyanarayan, T.S. (1996). Field and laboratory studies on selected ecological and behavioral aspects of sibling species of the *Anopheles culicifacies* complex, Ph.D. thesis, Delhi University, Delhi.
Sharma, D. (2004). Thar desert: Sitting on the tip of a malarial iceberg. *The Lancet*, **4,** 322.
Sharma, V.P. (1987). Community-based malaria control in India. *Parasitol. Today*, **3**(7), 222–226.
Sharma, V.P. (1988). Community-based bioenvironmental control of malaria in India. *Ann. Natl. Acad. Med. Sci, (India)*, **24**(3), 157–169.
Sharma, V.P. (1991). In G.A.T. Targett (ed), *Malaria Waiting for the Vaccine*. Wiley. pp. 49–66.
Sharma, V.P. (1996). Reemergence of malaria in India. *Ind. J.. Med. Res*., **103,** 26–45.
Sharma, V.P. (1998). Fighting malaria in India. *Curr. Sci*., **75**(11), 1127–1140.
Sharma, V.P. (1998). Getting the community involved. *World Health* 51st. Year No. 3, May–June. pp. 14–15.
Sharma, V.P. (1999). Current scenario of malaria in India. *Parassitologia*, **41,** 349–353.
Sharma, V.P. (2002). The contextual determinants of malaria. In E.A. Casman and H. Dowlatabadi (eds), *Resources for the Future*. Washington, USA.
Sharma, V.P. (2003). DDT: The fallen angel. *Curr. Sci*., **85**(11), 1532–1537.
Sharma, V.P. (2003). Malaria and poverty in India. *Curr. Sci*., **84**(4), 513–515.
Sharma, V.P. and Mehrotra, K.N. (1982a). Malaria resurgence. *Nature* (*Lond.*), **300,** 212.
Sharma, V.P. and Mehrotra, K.N. (1982b). Return of malaria. *Nature* (*Lond.*), **298,** 210.
Sharma, V.P. and Mehrotra, K.N. (1986). Malaria resurgence in India: A critical study. *Soc. Sci. Med*., **22,** 835–845.
Sharma, V.P. and Sharma, R.C. (1989). Community-based bioenvironmental control of malaria in Kheda district, Gujarat, India. *J. Am. Mosq. Contr. Assoc*., **5,** 514–521.

Sharma, V.P., Chandrahas, R.K., Ansari, M.A., Srivastava, P.K., Razdan, R.K., Batra, C.P., Raghavendra, R., Nagpal, B.N., Bhalla, S.C., and Sharma, G.K. (1986). Impact of DDT and HCH spraying on malaria transmission in villages with DDT and HCH resistant *Anopheles culicifacies*. *Ind. J. Malariol*, **23,** 27–38.

Sharma, V.P., Uprety, H.C., Nanda, N., Raina, V.K., Parida, S.K., and Gupta, V.K. (1982). Impact of DDT spraying on malaria transmission with resistant *Anopheles culicifacie*. *Ind. J. Malariol.*, **23,** 5–12.

Sharma, Y.D. (1995). Malaria menace. *Nature*, **373,** 279.

Singh, N. and Sharma, V.P. (2002). Patterns of rainfall and malaria in Madhya Pradesh, central India. *Ann. Trop. Med. Parasitol.*, **96**(4), 349–359.

Singh, O.P., Raghavendra, K., Nanda, N., Mittal, P.K., and Subbarao, S.K. (2002). Pyrethroid resistance in *Anopheles culicifacies* in Surat district, Gujarat, West India. *Curr. Sci.*, **82**(5), 547–550.

Subbarao, S.K. (1988). The *Anopheles culicifacies* complex and control of malaria. *Parasitology Today*, **4**(3), 72–75.

Subbarao, S.K. and Sharma, V.P. (1994). In: Sushil Kumar, A.K. Sen, G.P. Dutta, and R.N. Sharma (eds), *Tropical Diseases, Molecular Biology and Control Strategies*. CSIR-Publications and Information Directorate, New Delhi. pp. 377–386.

Subbarao, S.K. and Sharma, V.P. (1997). Anopheline species complexes and malaria control. *Ind. J. Med. Res.*, **106,** 164–173.

Subbarao, S.K., Adak, T., and Sharma, V.P. (1980). *Anopheles culicifacies*: Sibling species distribution and vector incrimination studies. *J. Com. Dis.*, **12,** 102–104.

Subbarao, S.K., Nanda, N. Vasantha, K., Dua, V.K., Malhotra, M.S., Yadav, R.S., and Sharma, V.P. (1994). Cytogenetic evidence for three sibling species in *Anopheles fluviatilis* (Diptera: Culicidae). *Ann. Ent. Soc. Am.*, **87**(1), 116–121.

Subbarao, S.K., Vasantha, K., Adak, T., and Sharma, V.P. (1983). *Anopheles culicifacies* complex: Evidence for a new sibling species, Species C. *Ann. Ent. Soc. Am.*, **76**(6), 985–988.

Subbarao, S.K., Vasantha, K., Adak, T., and Sharma, V.P. (1987). Seasonal prevalence of sibling species A and B of the taxon *Anopheles culicifacies* in villages around Delhi. *Ind. J. Malariol.*, **24,** 9–15.

Subbarao, S.K., Vasantha, K., Joshi, H., Raghvendra, K., Usha Devi, C., Satyanarayan, T.S., Cochrane, A., Nussenzweig, R.S., and Sharma, V.P. (1992). Role of *Anopheles culicifacies* sibling species in malaria transmission in Madhya Pradesh state, India. *Trans. R. Soc. Trop. Med. Hyg.*, **86,** 613–614.

Subbarao, S.K., Vasantha, K., Raghvendra, K., Sharma, V.P., and Sharma, G.K. (1988). *Anopheles culicifacies*: Sibling species composition and its relationship to malaria incidence. *J. Am. Mosq. Contr. Assoc.*, **4**(1), 29–33.

Suguna, S.G., Tewari, S.C., Mani, T.R., Hariyan, J., and Reuben, R. (1983). *Anopheles culicifacies* species complex in Thenpennaiyar river tract, Tamil Nadu. *Ind. J. Med. Res.*, **77,** 455–459.

Surendran, S.N., Abhayawardana, T.A., De Silva, B.G.D.N.K., Ramasamy, R., and Ramasamy, M.S. (2000). *Anopheles culicifacies* Y-chromosome dimorphism indicates sibling species (B and E) with different malaria vector potential in Sri Lanka. *Med. Vet. Entomol.*, **14,** 437–440.

Tewari, S.N., Prakash, A., Subbarao, S.K., Roy, A., Joshi, H., and Sharma, V.P. (1994). Correlation of malaria endemicity and *Anopheles culicifacies* sibling species composition and malaria antibody profile in district Allahabad (UP). *Ind. J. Malariol.*, **31,** 48–56.

Vasantha, K., Subbarao, S.K., and Sharma, V.P. (1991). *Anopheles culicifacies* complex: Population genetic evidence for species D (Diptera: Culicidae). *Ann. Ent. Soc. Am.*, **84**(5), 531–536.

Yadav, R.S., Sampath, T.R.R., and Sharma, V.P. (2001). Deltamethrin-treated bed nets for control of malaria transmitted by *Anopheles culicifacies* (Diptera: Culicidae) in India. *J. Med. Entomol.*, **38**(5), 613–622.

Zaim, M., Subbarao, S.K., Manouchehri, A.V., and Cochrane, A.H. (1993). Role of *Anopheles culicifacies S.l.* and *A. pulcherrimus* in malaria transmission in Ghassreghand (Baluchistan), Iran. *J. Am. Mosq. Contr. Assoc.*, **9,** 23–26.

Zavala, F., Gwadz, R.W., Collins, F.H., Nussenzweig, R.S., and Nussenzweig, V. (1982). Monoclonal antibodies to circumsporozoite proteins to identify the species of malaria parasite in infected mosquitoes. *Nature (Lond.)*, **299,** 737–738.

The Rate of Mutations of Human Genes*

J. B. S. Haldane

The rate at which genes mutate under natural conditions is a fundamental fact which is of great importance in any theory of evolution, and also for many practical problems. Such rates are known in *Drosophila* species, in *Zea mays*, and in bacteria, if we assume, as is probable, that we can legitimately speak of bacterial genes. They are not known in non-human vertebrates, because one cannot observe a million mice, let alone a million sheep or cows, of known parentage. They are however known in men, where large populations can be studied. While the most important work on this subject has been done in KEMP's institute at Copenhagen, I may perhaps be excused for speaking on it as PENROSE and I were the first to estimate human mutation rates, and the Danish workers have confirmed our admittedly rough estimates.

A human mutation rate can only be estimated under certain conditions. First, the gene must be detectable in the heterozygous or hemizygous condition. That is to say it must be a dominant like chondrodystrophy or a sex-linked recessive like haemophilia. Secondly, it must be very rare, lowering fitness so much that selection and mutation approximately balance one another. On the other hand it is not necessary, as might be thought, that the gene should be fully penetrant.

Let us see why these conditions are necessary. Suppose that we tried to estimate, from SJÖGREN's (1931) data, the mutation frequency of the lethal autosomal recessive gene for juvenile amaurotic idiocy. The frequency of homozygotes in Sweden is about 4×10^{-5}. If the gene were completely recessive, and if the population were in equilibrium we could say that as many genes were destroyed by selection as were produced by mutation in each generation, and the mutation rate would be 4×10^{-6}.

---

*Original publication: Haldane J. B. S. The rate of mutation of human genes. In: Proceedings of the Eighth International Congress of Genetics. (Issued as a suplimentory volume of HEREDITAS), pages 267-273, Lund, Sweden, 1949.

J. B. S. Haldane • University College, London.
Malaria: Genetic and Evolutionary Aspects, edited by Krishna R. Dronamraju and Paolo Arese, Springer, New York, 2006.

But we know that the population is not in equilibrium. As a result of urbanisation rural isolates have been broken up, and inbreeding diminished in the last century. The frequency of the condition was therefore probably a good deal higher 100 years ago than now. Nor do we know that it is fully recessive. If the heterozygotes, who are at least 300 times as common as homozygotes, have a fitness even 0.4 % greater than the average of the population, the gene will increase in frequency if there is no mutation at all. We cannot exclude such a possibility.

Again suppose we wished to measure the frequency with which the blood group gene *B* appears by mutation; if this frequency is as great as the greatest value so far recorded, namely 4×10^{-11}, then nearly one member of groups *B* and *AB* in 10,000 is due to mutation. We might hope to find one such case after investigating 10,000 children of these groups and their parents. There is no doubt however that technical errors and illegitimacy would give a far higher apparent mutation rate than this.

The best estimate of a human mutation rate is that for haemophilia, but MØRCH's (1941) work on chondrodystrophy (or achondroplasia) illustrates the three types of methods which may be used. This condition appears to be due to a dominant autosomal gene, or perhaps a deficiency. The homozygote is unknown, but is very probably lethal. Unfortunately the data on heredity are far from adequate. Older work suggests that the gene may occasionally be impenetrant. MØRCH's Danish data show 17 normal, 10 chondrodystrophic children of chondrodystrophics. The probability of so large a divergence from equality is .24. But one Swedish male described in his monograph had 11 normal children followed by 3 chondrodystrophics. There is only one chance in 35 that an ordinary gene would give such a ratio as 28 to 13, apart from the peculiar distribution in time. Perhaps chondrodystrophy is due to the loss of a chromosome, or to an extra chromosome or fragment which might be expected to segregate abnormally. It is conceivable that some chondrodystrophic zygotes are eliminated before birth. If so the mutation rate is rather higher than that here calculated.

MØRCH estimated the mutation frequency directly. Of 128,763 babies born in two hospitals, 14 were chondrodystrophic, but of these 3 had a chondrodystrophic parent. Thus the mutation rate μ per chromosome is $\frac{11}{2 \times 128,760}$ or $4.3 \times 10^{-6} \pm 1.3 \times 10^{-5}$. MØRCH found that the fitness of chondrodystrophics (as measured by the number of their children) was 19.6 % of that of their normal sibs. The frequency x at equilibrium is given by $x = \frac{2\mu}{1-f^{1}}$, or $\mu = \frac{1}{2}(1-f)$. x is not accurately known owing to the high infantile death rate. But on MØRCH's figures $x = 1.07 \times 10^{-4}$, so $\mu = 4.5 \times 10^{-5}$. Finally we may make a further calculation. Of MØRCH's

108 cases 98 were mutants, 10 had a chrondrodystrophic parent. Thus $\frac{98}{108}$ of all the dwarfs born were mutants. If $x = 1.07 \times 10^{-4}$ this gives $\mu = 4.9 \times 10^{-5}$. The weakest point in these latter two calculations is that though the frequency of chondrodystrophics in the living population is accurately known, that at birth is not. And as about 80 % die in their first year, this may introduce a serious inaccuracy. Nevertheless the different results agree very well.

MØRCH believes the gene to be fully penetrant. It is however worth showing that the calculated mutation rate would not be much altered if it had a penstrance of anything over 50 %. In a stationary population, provided the gene does not greatly alter viability unless it is expressed, the penetrance is unimportant. As many genes appear by mutation as are destroyed by selection. And provided the fitness is known to be small the mutation rate can be calculated as before.

In the case of haemophilia the genetics are much clearer. But we cannot yet get a direct estimate of the mutation frequency. It would be possible to explain the persistence of haemophilia, despite its selective disadvantage, if segregation were biassed, and heterozygous women gave an excess of haemophilie sons and heterozygous daughters. HALDANE and PHILIP (1939) and HALDANE (1947) showed that this was not so. When allowance is made for bias in the pedigrees due to the fact that mothers and other ancestresses of propositi (probands) are necessarily heterozygous apart from mutation, segregation is in accordance with theory. 47.1 ± 6.1 % of the sons and 52.7 ± 5.9 % of the daughters of heterozygotes received the gene for haemophilia. There are also numerous pedigrees only explicable by mutation. In this case if x be the frequency in males at birth, f the fitness, and μ the mutation rate, $\mu = \frac{1}{0}(1-f)x$. ANDREASSEN (1943) found 81 haemophilies among 1,620,000 Danish males, and on the basis of their mortality table estimates the frequency at birth to be three times as great, so that $x = 1.33 \times 10^{-6}$. His estimate of f was however based not on the progeny of his propositi, but on that of all haemophilics in his pedigrees, who naturally included a number of unusually fertile haemophilies. Thus I think his estimate of .504 for f is far too high. Since half of all Danish haemophilies die by the age of 18, it would imply that those who survived, in spite of the fact that they often die during the reproductive period, were more fertile than the average. The true value of f is about, 280, giving $\mu = 3.16 \times 10^{-5}$.

The mutation rates for retinoblastoma (PHILIP and SORSBY, unpublished), apiridia (MØLLENBACH, 1947) and epiloja (GUNTHER and PENROSE, 1935) are obtained by essentially similar methods. The gene for retinoblastoma is not completely penetrant, but this makes very little difference to the calculation. For when a gene is nearly lethal, that is to say f is nearly zero, small errors in the estimation of f do not greatly affect the value of $(1 - f)x$. It may be added that for retinoblastoma f seems to be about -2. It is tending to rise as a result of successful operations, but to fall as a result of eugenie advice. For aniridia, 21 of MØLLENBACH's 40 cases were mutants, so f is about 0.5.

PÄTAU and NACHTSHEIM (1946) made a rough estimate of the mutation frequency of the dominant Pelger anomaly of the leucocytes as follows. They examined the parents of 12 patients, and found that in 2 cases both were normal. The true frequency is probably less, because if both parents are normal they are more likely to survive than if one has the anomaly. If a fraction y of all the Pelger genes arise by mutation, and if its frequency at birth is x (assuming no pre-natal elimination) then $f = 1 - y$, and $\mu = \frac{1}{2}xy$. Here x is taken to be about 10^{-5}, so μ is about 8×10^{-5}, in agreement with the other values. But both x and y are at present very roughly estimated. It should be quite possible to improve these values. In such a case however it will be essential to establish that the gene has 100 % penetrance.

Finally NEEL and VALENTINE (1947) have calculated a mutation rate for the gene which when homozygous gives rise to the lethal condition of thalassemia major (Cooley's anaemia) while the heterozygote has a mild microcytic anaemia with decreased fragility of the erythrocytes. The gene is found in about 4.1 % of the people of Italian origin in Rochester, N.Y. If the heterozygote had normal viability, equilibrium would be secured by a mutation rate of 4×10^{-4}. On the other hand if the heterozygote had an increased fitness of only 2.1 % there would be equilibrium without any mutation at all.

NEEL and VALENTINE believe that the heterozygote is less fit than normal, and think that the mutation rate is above 4×10^{-4} rather than below it. I believe that the possibility that the heterozygote is fitter than the normal must be seriously considered. Such increased fitness is found in the case of several lethal and sublethal genes in Drosophila and Zea. A possible mechanism is as follows. The corpuscles of the annemic heterozygotes are smaller than normal, and more resistant to hypotonic solutions. It is at least conceivable that they are also more resistant to attacks by the sporozoa which cause malaria, a disease prevalent in Italy, Sicily, and Greece, where the gene is frequent. Similarly, the gene, which causes an anaemia similar to that of iron deficiency, might be harmless or even useful to persons on an iron deficient diet, though causing a relative anaemia when the diet is more generous. Numerous other similar hypotheses could be framed. On many of them, the gene would lower fitness in America, but might, at least in the heterozygous *condition, have the opposite effect in Greece or Sicily. Until more is known of the physiology of this gene in various environments I doubt if we can accept the hypothesis that it arises very frequently by mutation in a small section of the human species.*

The other mutation rates, which seem to be better established, range between 4×10^{-5} and 4×10^{-6}. ANDREASSEN's data however seem to show that haemophilic families generally start with a heterozygous female, that is to say the mutation rate is higher in male than in female *X*-chromosomes, perhaps 10 or more times higher. The tests now available for helerozygosily in women should clear this point up, and further work is needed on the apparently large increase in mutation with age found by MØRCH. I suggest that, for the sake of uniformity, mutation rates should always be stated per chromosome per generation. In the Danish data the rate is sometimes stated in this way, and sometimes as the fraction of mutant individuals among all births, which is of course twice the above number in the case of dominants.

In the case of the blood group genes an upper limit to the mutation rate in women can be found. ANDREASSEN (1947) has examined 8,000 mothers of type *O* and their children. There were no exceptions, that is to say children of group *AB*. As the frequencies of the *A* and *B* genes in Denmark are 28 % and 7 % we can say that *O* had not mutated to *A* in about 560 cases nor to *B* in 2,240 cases. Similarly no children of 800 *AB* mothers were of group *O*, though in about 520 cases this could have been detected. Altogether there are about 6,000 such cases in the literature, which include one almost certain mutation, HASELHORST and LAUER's (1930, 1931) case of an *AB* mother with an *O* child. As this child was defective in several respects it possibly lacked a chromosome or part of one, though some of the defects may have been due to birth injury. Thus at this locus there has been one chromosomal aberration or one point mutation, very possibly the former, in some 6,000 onses. Even if the blood group genes are as mutable as the normal allelomorph of haemophilia, we should have to examine over five times as many mother—child pairs before we were likely to find another point mutation. The figures for the *M* and *N* locus are of the same order. But in both cases it is possible that the mutation rate is higher in men than in women. The methods cited above would not disclose such a rate.

The rates here reported are presumably those of particularly mutable human genes. They are about 10 times as high as those of the most mutable of the normal genes in *Drosophila melanogaster*, though the cut and yellow loci in *D. subobscura* seem to have higher rates. That is to say they are higher per generation, or per nuclear division, since there are

Table 1

Gene	Mutation Rate
Chondrodystrophia	4×10^{-5}
Haemophilia	3×10^{-5}
Relinoblastoma	1.4×10^{-5}
Aniridia	$> 1.2 \times 10^{-5}$
Epiloia	$4—8 \times 10^{-5}$
Polger anomaly	8×10^{-5}?
Thalassemia	$> 4 \times 10^{-1}$?

only about twice as many nuclear divisions per generation in man as in *Drosophila*. But the daily rates are about 500 times as high in *Drosophila* as in man. Now MULLER (1930) and MOTT-SMITH calculated that natural radiations and high-speed particles would only account for about one mutation per thousand in *Drosophila*. It follows that, if the radiosensitivity of human genes is about the same as in *Drosophila*, radiation may account for all, or almost all, human sublethal mutations. However the fact that the genes in *Sciara* are highly insensitive to radiation suggests the need for caution before reaching such a conclusion.

I am aware that the arguments in this paper are open to several criticisms. To mention only one, they were propounded under the influence of the philosophy to which our president referred in somewhat unfavourable terms. Nevertheless I hope I have shown that man provides unique material for the study of mutation, provided that special statistical methods are employed, and above all that care is taken to avoid the rather numerous fallacies which beset human genetics, and which are particularly liable to deceive workers whose previous experience has lain in the experimental study of plant and animal genetics.

References

1. ANDREASSEN, M. 1943. Haemofili i Danmark.—Op. ex. dom. biol. hered. hum. Univ. Hamlensis, 6 (Copenhagen).
2. ANDRESEN. P. H. 1947.—Aeta Path. 24, 546–558.
3. GUNTHEE, E. R. and PENROSE, L. S. 1935.—J. Genet. 31, 420–280.
4. HALDANE, J. B. S. 1947.—Ann. Eugen. 13. 252–271.
5. HALDANE, J. B. S. and PHILIP, U. 1939.—J. Genel. 38, 193–900.
6. HASHLHORST, G. and LAUEA, A. (1930, 1931).—Zeitschr. f. dic ges. Anat. 15, 205: 16, 227.
7. MULLER, H. J. 1930.—Am. Nat. 84, 220.
8. MØLLENBACH, C. J. 1947. Medfodie defekter j øjets indre hinder, Kilnik og arvelighedfurhold.—Op. ex dom. biol. hered. hum. Univ. Faxfnieusis, 15 (Copenhagen).
9. MORCH, T. 1941. Chondrodystrophic dwarfs in Denmark.—Copenhagen, Ejnar Munksgaard.
10. NEEL, J. V. and VALENTINE, W. N. 1947.—Geneties 32, 98—63.
11. PÄTAU, K. and NACHTSHEIM, H. 1946.—Z. Naturforsch. 1, 345.
12. SIÖCREN, T. 1931.—Ilureditas, XIV, 107–125.

Disease and Evolution*

J. B. S. Haldane

RIASSUNTO. – Negli esempi addotti dai biologi per mostrare come la selezione naturale agisce, la struttura o la funzione presa in esame è per lo più collegata con la protezione contro forze naturali a vverse, contro predatori, oppure con la conquista di alimento o dell'altro sesso. L'A. mostra che la lotta contro le malattie, e in particolare contro le malattie infettive, ha rappresentato un fattore evolutivo molto importante e che alcuni dei suoi risultati sono diversi da quelli raggiunti attraverso le forme consuete della lotta per l'esistenza.

RÉSUMÈ. – Les exemples portés par les biologistes pour montrer comment la séléction naturelle opère tiennent compte d'ordinaire de structures ou de fonctions liées à la protection contre des forces naturelles hostiles, contre des prédateurs, ou bien liées à la conquéte de la nourriture ou du sexe opposé. L'A. montre que la lutte contre les maladies, et en particulier contre les maladies infectieuses, a représenté un facteur évolutif très important et que qualques-uns parmi ses résultats diffèrent bien de ceux qui ont été atteints par les formes ordinaires de la lutte pour la vie.

SUMMARY. – Examples quoted by biologists, in order to show how natural selection is working, almost present structures or functions concerned either with protection against natural forces or against predators, or with purchase of food or mates. The Author suggests that the struggle against diseases, and especially infectious diseases, has been a very important evolutionary agent and that some of its results have been rather unlike those of the struggle for life in its common meaning.

It is generally believed by biologists that natural selection has played an important part in evolution. When however an attempt is made to show how natural selection acts, the structure or function considered is almost always one concerned either with protection against natural «forces» such as cold or against predators, or one which helps the organism to obtain

---

*Original publication: Haldane, J. B. S. (1949). Disease and evolution. Supplement to La Ricerca Scientifica 19:68–76.

J. B. S. Haldane • University College, London.
Malaria: Genetic and Evolutionary Aspects, edited by Krishna R. Dronamraju and Paolo Arese, Springer, New York, 2006.

food or mates. I want to suggest that the struggle against disease, and particularly infectious disease, has been a very important evolutionary agent, and that some of its results have been rather unlike those of the struggle against natural forces, hunger, and predators, or with members of the same species.

Under the heading infectious disease I shall include, when considering animals, all attacks by smaller organisms, including bacteria, viruses, fungi, protozoa, and metazoan parasites. In the case of plants it is not so clear whether we should regard aphids or caterpillars as a disease. Similarly there is every gradation between diseases due to a deficiency of some particular food constituent and general starvation.

The first question which we should ask is this. How important is disease as a killing agent in nature? On the one hand what fraction of members of a species die of disease before reaching maturity? On the other, how far does disease reduce the fertility of those members which reach maturity? Clearly the answer will be very different in different cases. A marine species producing millions of small eggs with planktonic larvae will mainly be eaten by predators. One which is protected against predators will lose a larger proportion from disease.

There is however, a general fact which shows how important infectious disease must be. In every species at least one of the factors which kills it or lowers its fertility must increase in efficiency as the species becomes denser. Otherwise the species, if it increased at all, would increase without limit. A predator cannot in general be such a factor, since predators are usually larger than their prey, and breed more slowly. Thus if the numbers of mice increase, those of their large enemies, such as owls, will increase more slowly. Of course the density-dependent check may be lack of food or space. Lack of space is certainly effective on dominant species such as forest trees or animals like *Mytilus*. Competition for food by the same species is a limiting factor in a few phytophagous animals such as defoiiating caterpillars, and in very stenophagous animals such as many parasitoids. I believe however that the density-dependent limiting factor is more often a parasite whose incidence is disproportionately raised by overcrowding.

As an example of the kind of analysis which we need, I take Varley's (1947) remarkable study on *Urophora jaceana*, which forms galls on the composite *Centaurea nigra*. In the year considered 0.5% of the eggs survived to produce a mature female. How were the numbers reduced to $\frac{1}{200}$ of the initial value?

If we put $200 = e^k$, we can compare the different killing powers of various environmental agents, writing $K = k_2 \div k_2 \div k_5 \div \ldots$, where k_r is a measure of the killing power of each of them. The result is given in Table 1. Surprisingly, the main killers appear to be mice and voles (*Mus*,

Table 1

Month	Density per square metre	Cause of death	k.
July	203,0	—	—
,,	184,7	Infertile eggs.	0,095
,,	147,6	Failure to form gall.	0,224
,,	144,6	? Disease.	0,021
,,	78,8	*Eurytoma curla*	0,607
Aug., Sept.		Other parasitoids.	0,222
,,	50,2	Caterpillars.	0,234
Winter	19,2	Disappearance, probably mice.	0,957
,,	7,0	Mice.	1,000
,,	5,2	Unknown	0,297
May-June		Birds.	0,000
,,	3,6	Parasitoids	0,270
July	2.03	Floods.	0,581
			4,606

Microtus, etc.) which eat the falien gails and account for at least 22%, and perhaps 43% of *k*. Parasitoids account for 31 % of the total kill, and the effect of *Eurytoma curla* was shown to be strongly dependent on host density, and probably to be the main factor in controlling the numbers of the species, since the food plants were never fully occupied.

When we have similar tables for a dozen species we shall know something about the intensity of possible selective agencies. Of course in the case of *Urophora jaceana* analysis is greatly simplified by the fact that the imaginal period is about 2 % of the whole life cycle, so that mortality during it is unimportant.

A disease may be an advantage or a disadvantage to a species in competition with others. It is obvious that it can be a disadvantage. Let us consider an ecological niche which has recently been opened, that of laboratories where the genetics of small insects are studied. A number of species of *Drosophila* are well adapted for this situation. Stalker attempted to breed the related genus *Scaplomyza* under similar conditions, and found that his cultures died of bacterial disease. Clearly the immunity of *Drosophila* to such diseases must be of value to it in nature also.

Now let us take an example where disease is an advantage. Most, if not all, of the South African artiodaetyls are infested by trypanosomes such as *T. rhodesiense* which are transmitted by species of *Glossina* to other mammals and, sometimes at least, to men. It is impossible to introduce a species such as *Bos taurus* into an area where this infection is prevalent. Clearly these ungulates have a very powerful defence against invaders. The latter may ultimately acquire immunity by natural selection, but this is a very slow process, as is shown by the fact that the races of cattle belonging to the native African peoples have not yet acquired it after some centuries of sporadie exposure to the infection. Probably some of the wild ungulates die of, or have their health lowered by the trypanosomes, but this is a small price to pay for protection from other species.

A non-specific parasite to which partial immunity has been acquired, is a powerful competitive weapon. Europeans have used their genetic resistance to such viruses as that of measles (rubeola) as a weapon against primitive peoples as effective as fire-arms. The latter have responded with a variety of diseases to which they are resistant. It is entirely possible that great and, if I may say so, tragic episodes in evolutionary history such as the extinction of the Noto-ungulata and Litopterna may have been due to infectious diseases carried by invaders such as the ungulates, rather than to superior skeletal or visceral developments of the latter.

A suitable helminth parasite may also prove a more efficient protection against predators than horns or cryptic coloration, though until much more is known as to the power of helminths in killing vertebrates or reducing their fertility, this must remain speculative.

However it may be said that the capacity for harbouring a non-specific parasite without grave disadvantage will often aid a species in the struggle for existence. An ungulate species which is not completely immune to *Trypanosoma rhodesiense* has probably (or had until men discovered the life history of this parasite) a greater chance of survival than one which does not harbour it, even though it causes some mortality directly or indirectly.

Mortality of *Urophora jaceana* in 1935-1946, after Varley.

The winter disappearance was probably due to galls carried off by mice. The mortality attributed to mice is based on counts of galls bitten open. The k due to *Euryloma curla* in the preceding year was. 0,069, in the subsequent year 0,137. This cause of death depends very strongly on host and parasitoid densities. The caterpillars killed the larvae by eating the galls.

I now pass to the probably much larger group of cases where the presence of a disease is disadvantageous to the host. And here a very elementary fact must be stressed. In all species investigated the genetical diversity as regards resistance to disease is vastly greater than that as regards resistance to predators.

Within a species of plant we can generally find individuals resistant to any particular race of rust (Uredineae) or any particular bacterial disease. Quite often this resistance is determined by a single pair, or a very few pairs, of genes. In the same way there are large differences between different breeds of mice and poultry in resistance to a variety of bacterial and virus diseases. To put the matter rather figuratively, it is much easier for a mouse to get a set of genes which enable it to resist *Bacillus typhi murium* than a set which enable it to resist cats. The genes commonly segregating in plants have much more effect on their resistance to small animals which may be regarded as parasites, than to larger ones. Thus a semiglabrous mutant of *Primula sinensis* was constantly infested by aphids, which however are never found on the normal plant. I suppose thornless mutants of *Rubus* are less resistant to browsing mammals than the normal type, but such variations are rarer.

Anyone with any experience of plant diseases will of course point out that the resistance of which I have spoken is rarely very general. When a variety of wheat has been selected which is immune to all the strains of *Puccinia graminis* in its neighbourhood, a new strain to which it is susceptible usually appears within a few years, whether by mutation, gene recombination, or migration. Doubtless the same is true for bacterial and virus diseases. The microscopic and sub-microscopic parasites can evolve so much more rapidly than their hosts that the latter have little chance of evolving complete immunity to them. It is very remarkable that *Drosophila* is as generally immune as it is. I venture to fear that some bacillus or virus may yet find a suitable niche in the highly overcrowded *Drosophila* populations of our laboratories, and that if so this genus will lose its proud position as a laboratory animal. The most that the average species can achieve is to dodge its minute enemies by constantly producing new genotypes, as the agronomists are constantly producing new rust-resistant wheat varieties.

Probably a very small biochemical change will give a host species a substantial degree of resistance to a highly adapted microorganism. This has an important evolutionary effect. It means that it is an advantage to the individual to possess a rare biochemical phenotype. For just because of its rarity it will be resistant to diseases which attack the majority of its fellows. And it means that it is an advantage to a species to be biochemically diverse, and even to be mutable as regards genes concerned in disease resistance. For the biochemically diverse species will contain at least some members capable of resisting any particular pestiience. And the biochemically mutable species will not remain in a condition where it is resistant to all the diseases so far encountered, but an easy prey to the next one. A beautiful example of the danger of homogeneity is the case of the cuitivated banana clone « Gross Michel » which is well adapted for export and has been widely planted in the West Indies. However it is susceptible to a root infection by the fungus *Fusarium cubense* to which many varieties are immune, and its exclusive cultivation in many areas has therefore had serious economic effects.

Now every species of mammal and bird so far investigated has shown a quite surprising biochemical diversity revealed by serological tests. The antigens concerned seem to be proteins to which polysaccharide groups are attached. We do not know their functions in the organism, though some of them seem to be part of the structure of cell membranes. I wish to suggest that they may play a part in disease resistance, a particular race of bacteria or virus being adapted to individuals of a certain range of biochemical constitution, while those of other constitutions are relatively resistant. I am quite aware that attempts to show that persons of a particular blood group are specially susceptible to a particular disease have so far failed. This is perhaps to be expected, as a disease such as diphtheria or tuberculosis is caused by a number of biochemically different races of pathogens. The kind of investigation needed is this. In a particular epidemic, say of diphtheria, are those who are infected (or perhaps those who are worst affected) predominantly drawn from one serological type (for example *AB, MM*, or *BMM*)? In a different epidemic a different type would be affected.

In addition, if my hypothesis is correct, it would be advantageous for a species if the genes for such biochemical diversity were particularly mutable, provided that this could be achieved without increasing the mutability of other genes whose mutation would give lethal or sublethal genotypes. Dr. P. A. Gorer informs me that there is reason to think that genes of this type are particularly mutable in mice. Many pure lines of mice have split up into sublines which differ in their resistance to tumour implantation. This can only be due to mutation. The number of loci concerned is comparable, it would seem, with the number concerned with coat colour. But if so their mutation frequency must be markedly greater.

We have here, then, a mechanism which favours polymorphism, because it gives a selective value to a genotype so long as it is rare. Such mechanisms are not very common. Among others which do so are a system of self-sterility genes of the *Nicotiana* type. Here a new and rare gene will always be favoured because pollen tubes carrying it will be able to grow in the styles of all plants in which it is absent, while common genes will more frequently meet their like. However this selection will only act on genes at one locus, or more rarely at two or three. A more generally important mechanism is that where a heterozygote is fitter than either homozygote, as in *Paratettix texanus* (Fisher 1939) and *Drosophila pseudoobscura* (Dobzhansky 1947). This does not, however, give an advantage to rarity as such. It need hardly be pointed out that, in the majority of cases where it has been studied, natural selection reduces variance.

I wish to suggest that the selection of rare biochemical genotypes has been an important agent not only in keeping species variable, but also in speciation. We know, from the example of the *Rh* locus in man, that biochemical differentiation of this type may lower the effective fertility of matings between different genotypes in mammals. Wherever a father can induce immunity reactions in a mother the same is likely to be the case. If I am right, under the pressure of disease every species will pursue a more or less random path of biochemical evolution. Antigens originally universal will disappear because a pathogen had become adapted to hosts carrying them, and be replaced by a new set, not intrinsically more valuable, but favouring resistance to that particular pathogen. Once a pair of races is geographically separated they will be exposed to different pathogens. Such races will tend to diverge antigenically, and some of this divergence may lower the fertility of crosses. It is very striking that Irwin (1947) finds that related, and still crossable, species of *Columba, Streplopelia*, and allied genera differ in respect of large numbers of antigens. I am quite aware that random mutation would ultimately have the same effect. But once we have a mechanism which gives a mutant gene as such an advantage, even if it be only an advantage of one per thousand, the process will be enormously accelerated, particularly in large populations.

There is still another way in which parasitism may favour speciation. Consider an insect in which a parasitoid, say an ichneumon fly, lays its eggs. And let us suppose both host and parasite to have an annual cycle, the parasite being specific to that particular host. To simplify matters still further, we shall suppose that the parasite is the only density-dependent factor limiting the growth of the host population, whilst the density-dependent factor limiting the growth of the parasite population is the difficulty of finding hosts. It is further assumed that the parasitoid only lays one egg in each host, or that only one develops if several are laid. Varley showed that all these assumptions are roughly true for *Urophora jaceana* and *Eurytoma curta*.

Let e^k be the effective fertility of the host, that is to say let k be the mean value of the natural logarithm of the number of female-producing eggs laid per female. Let k_1 be the killing power of agents killing during the part of the life cycle before the host is not infested. Let k_2 be the killing power of agents other than the parasite killing during that part of the life cycle when it is infested, and therefore killing the same fraction $1-e^{-k_2}$ of parasites as of hosts. Let k_3 be the killing power of agents killing after the parasites have emerged as imagines.

Let x be the equilibrium density of adult hosts, y that of parasitoids, and a the mean area of search of the parasitoid.

Then the fraction of hosts which are not parasitized must be e^{-k_4}, where $k_1+k_2+k_3+k_4=K$: Nicholson and Bailey (1933) showed that $k_4=ay$. But the host density available for parasitism is e^{K-k_1}. Of these $(1-e^{-k_4})$ e^{K-k_1} are parasitized. And $e^{K-k_1-k_2}$ $(1-e^{-k_4})$ of the parasites live to emerge. This is diminished by a factor e^{-k_1}. The equilibrium densities are given by

$$x = \frac{e^{k_1 - k_3}\, y}{e^{K - k_1 - k_2 - k_3} - 1},$$

$$y = \frac{K^{k_1 - k_2 - k_3}}{a},$$

though there is perpetual oscillation round this equilibrium. Now consider the effect of changes in these parameters. Any gene in the host which increases $K - k_1 - k_2 - k_3$ will give its carriers a selective advantage over their fellows, and will therefore spread through the population. This will cause an increase in the density of parasitoids. If it acts before or during parasitisation, thus diminishing k_1 or k_2, it will diminish the equilibrium value of x. If it acts after parasitisation is over, thus diminishing k_3, it will increase the equilibrium value of x, though not very much. But since we have supposed that the parasitoids only emerge shortly before the end of the hosts life cycle, every increase in its adaptation to environmental factors other than the specific parasite will diminish the numbers of adult hosts, though it may increase the number of their larvae at an early stage. This is a striking example of the way in which the survival of the fittest can make a species less fit.

A concrete case would be a gene which, by improving cryptic or aposematic coloration of the larvae, enabled more of them to escape predators, and therefore more parasites to do so. Since the host population is denser they will parasitise a larger fraction of the hosts and thus reduce their number. Since a larger fraction of parasites escape, equilibrium will be reached with a lower host density. Fewer of the caterpillars «nati a far l'angelica farfalla» will achieve this end. More of them will give rise to ichneumons or chalcids.

Natural selection will also favour genes which enable the bost to resist the parasitoid, but the latter will also increase its efficiency by natural selection. As Nicholson and Bailey showed, every increase in the area searched by it will diminish the density of both hosts and parasites.

The best that can be said for this tendency, from the host's point of view, is that it makes it less likely to become extinct as the result of other agencies. For the parasitoid being dependent on the density of the host population, will allow its numbers to increase rapidly after any temporary fall.

The host can hardly hope to throw off the parasite permanently by changing its life cycle, developing immunity (if this is possible) or otherwise. But it can reduce its numbers by speciating. For suppose that the pair of species is replaced by two host-parasitoid pairs, the population will be doubled, in so far as the parasitoid is the regulator. It is unlikely that a species can divide sympatrically, but the reduction in numbers caused by parasitoidism will leave food available for other immigrant species of similar habits, even if they are equally parasitised.

Thus certain types of parasitism will tend to encourage speciation, as others encourage polymorphism. This will specially be the case where the parasite is very highly adapted to its host, the most striking cases of adaptation being probably those of the parasitoid insects.

We see then that in certain circumstances, parasitism will be a factor promoting polymorphism and the formation of new species. And this evolution will in a sense be random. Thus any sufficiently large difference in the times of emergence or oviposition of two similar insect species will make it very difficult for the same parasitoid to attack both of them efficiently. So will any sufficiently large difference in their odours. We may have here a cause for some of the apparently unadaptive differences between related species.

Besides these random effects, disease will of course have others. It is clear that natural selection will favour the development of all kinds of mechanisms of resistance, including tough cuticles, phagocytes, the production of immune bodies, and so on. It will have other less obvious effects. It will be on the whole an antisocial agency. Disease will be less of a menace to animals living singly or in family groups than to those which live in large communities. Thus it is doubtful if all birds could survive amid the faecal contamination which characterises the colonies of many sea birds. A factor favouring dispersion will favour the development of methods of sexual recognition at great distances such as are found in some Lepidoptera.

Again, disease will set a premium on the finding of radically new habitats. When our ancestors left the water, they must have left many of their parasites behind them. A predator which ceases to feed on a particular prey, either through migration or changed habits, may shake off a cestode which depends on this feeding habit. When cerebral development has gone far enough to make this possible, it will favour a negative reaction to faecal odours and an objection to cannibalism, and will so far be of social value. A vast variety of apparently irrelevant habits and instincts may prove to have selective value as a means of avoiding disease.

A few words may be said on non-infectious diseases. These include congenital diseases due to lethal and sub-lethal genes. Since mutation seems to be non-specific as between harmful and neutral or beneficial genes, and mutation rate is to some extent inherited, it follows that natural selection will tend to lower the mutation rate, and this tendency may perhaps go so far as to slow down evolution. It will also tend to select other genes which neutralise the effect of mutants, and thus to make them recessive or even ineffective, as Fisher has pointed out. Whether the advantage thus given to polyploids is ever important, we do not know. But the evolution of dominance must tend to make the normal genes act more intensely and thus probably earlier in ontogeny, so that a character originally appearing late in the life cycle will tend to develop earlier as time goes on.

Deaths from old age are due to the breakdown of one organ or another, in fact to disease, and the study of the mouse has shown that senile diseases such as cancer and nephrosis are often congenital. In animals with a limited reproductive period senile disease does not lower the fitness of the individual, and increases that of the species. A small human community where every woman died of cancer at 55, would be more prosperous and fertile than one where this did not occur. Senile disease may be an advantage wherever the reproductive period is limited; and even where it is not, a genotype which lead to disease in the 10 % or so of individuals which live longest may be selected if it confers vigour on the majority. As Simpson (1944) has emphasized, some of the alleged cases of hypertely can be explained in this way.

Deficiency diseases are due to the lack of a particular food constituent which an organism or its symbionts cannot synthesize or make from larger molecules. They must act as a selective agent against the loss of synthetic capacity which is a very common type of mutation in simple organisms at least, and in favour of genotypes with a varied symbiotic flora. They might thus have speeded up the evolution of the ruminants, whose symbionts probably make vitamins as weir as simple nutrients like acetic acid. To be precise it might be an advantage to have a small rumen where symbionts made B vitamins before it got large enough to add appreciably to the available calories.

On the other hand Rudkin and Schulz (1948) have shown that deficiency diseases can select mutants which utilize the nutrilite in question less than does the normal type. In particular the vermilion mutant of *Drosophila melanogaster* does not form the brown eye pigment ommatin from tryptophane. It is more viable than the wild type on a diet seriously deficient in tryptophane. Thus deficiency diseases may cause a regressive type of evolution characterized by the loss of capacity to utilise rare nutrilites for synthesis.

In this brief communication I have no more than attempted to suggest some lines of thought. Many or all of them may prove to be sterile. Few of them can be followed profitably except on the basis of much field work.

References

Dobzhansky, T. 1947. Genetics of natural populations XIV. A response of certain gene arrangements in the third chromosome of *Drosophila subobscura* to natural selection. *Genetics*, 32: 142.

Fisher, R. A. 1939. Selective forces in wild populations of *Parateltix texanus*. *Ann. Eugen.*, 9: 109.

Irwin, M. R. 1949. Immunogenetics. *Advances in Genetics*, 1: 133.

Nicholson, I. J. and Bailey, V. A. 1935. The balance of animal populations Pt. 1. *Proc. Zool. Soc. London:* 551-598.

Rudkin, G. T. and Schulz, J. 1948. A comparison of the Tryptophane requirements of mutant and wild type *Drosophila melanogaster. Proc. Int. Cong. Genetics, Stockholm*. July 1948.

Simpson, G. G. 1944. *Tempo and mode in Evolution*. New York.

Varley, G. C. 1947. Natural control of population balance in the knapweed gallfly (*Urophora jaccana*). *Journ. Anim. Ecol.*, 16: 139.

Discussion

Montalenti. Sottolinea l'importanza delle vedute espresse dal Prof. Haldane. Ricorda il caso della microcitemia o talassemia, studiato da Silvestroni, Bianco e Montalenti. Qui un gene, letale allo stato omozigote (morbo di Cooley) si trova, allo stato eterozigote, con tale frequenza in alcune popolazioni (più del 10 %) che bisogna ammettere che esso rappresenli in questa condizione un vantaggio per gli individui che lo portano. Poiché da alcune ricerche, tutt'altro che complete, sembra che il gene sia più frequente in zone malariche, il Prof. Haldane ha suggerito in comanicazione verbale che gli individui microcitemici, i quali fra l'altro hanno resistenza globulare aumentata, possano essere più resistenti all'infezione malarica. Comunque è questo un caso interessante di eterosi, che si ricollega a quanto ha illustrato il Prof. Haldane.

Jucci. La relazione del Prof. Haldane ha sviluppato mágistralmente, in modo quanto mai súggestivo, un argomento del più alto interesse. Desidero fare qualche commento su qualcuno dei tanti aspetti del problema. Non conoscevo le ricerche di Varley su *Urophora*. Certo g i insetti gallicoli offrono un materiale particolarmente adatto. Anni fa comincial a raccog'iere dati analoghi sul cecidomide *Mikiola fagi*. Ma la mia attenzione fu particolarmente assorbita dal fatto che nella zona da me esplorata – ogni singolo faggio iungo la strada che saie al Terminillo, nel tratto da 1000 a 1200 metri di altezza – si presentava una varietà di comportamento spiccatissima. Accanto a un faggio carico di galle spesso un altro ne era del tutto sprovvisto, come per una variabilità genetica di comportamento della pianta ospite. Le ricerche verranno riprese ed approfondite ora che stiamo per organizzare una Stazione montana di Genetica sul Terminillo, all'altezza di 1700 m. Suggestive le considerazioni del Prof. Haldane sul pericolo che per la specie presenta una eccessiva omogeneità e sel vantaggip di una capacità a presentare mutazioni biochimiche in rapporto alla possibilità di esprimere dalla costituzione genetica razze resistenti a parassiti Io ho avuto occasione di studiare a lungo le diverse recettività di due bombici serigeni, *Phylosamia ricini* e *Ph. cynthia*, al virus del giallume. La *ricini*, forma domestica, è resistente: portatore sano; la *cynthia*, selvatica sull'ailanto, recettiva come il *Bombyx*. Forse è per questa ragione che in Italia la *cynthia* è diffusa largamente, ma ovunque poco abbondante. Una o poche mutazioni (nella F2 dell'incrocio si ha la disgiunzione dei fattori con ritorno a forme resistenti e recettive, come le parentali) possono aver determinato nella *ricini* la capacità ad essere allevate in massa. A proposito della tendenza antisociale che il periodo delle malattie infettive può imprimere a molte specie di organismi, noterò che per l'evoluzione degli insetti sociali deve certo avere avuto larga importanza l'acquisizione di forte resistenza alle infezioni. Sarebbe interessante paragonare a questo riguardo la recettività della forma domestica (*Apis mellifica*) e di forme affini selvatiche alla peste delle api e simili forme epidemiche. Una delle vie per le quali il fattore malattia da infestione o infezione deve avere profondamente influenzato il processo evolutivo è stata certo quella della simbiosi, che va considerata come uno stato di alleanza subentrato a un periodo più o meno lungo di guerra fra due organismi. Caratteristico il caso dei batteri simbionti nel tessuto adiposo di Blattidi e di Termiti. La simbiosi risale ai Protoblattoidi del Carbonifero, come lo dimostra l'identità dei processi di trasmissione dei batteri da una generazione all'altra nei Blattidi e nel *Mastotermes*. Gli Isotteri hanno lasciato cadere lungo la via filogenetica la simbiosi con i batteri forse perché hanno trovato assai più vantaggiosa la simbiosi con i flagellati dell'intestino che hanno loro permesso la conquista del mondo della cellulosa. Interessantissimo l'accenno del Prof. Haldane alle malattie di carenza: mutazioni in questo senso possono aver sollecitato lo stabilirsi di simbiosi nelle quali l'associato veniva ad offrire il fattore accessorio che l'ospite non era più capace di produrre e che non poteva trovare nell'ambiente.

HADORN. Bei Bakterien gibt es zahlreiche Beispiele, die zeigen dass biochemische Mutationen die zu synthetischen Defekten führen, gleichzeitig eine Resistenz gegen Infektionen (Phagen) bedingen. Vielleicht könnte dies also Modell dienen für den positiven Selektionswert im Falle von Thalassemia.

Wie kann man erklären, dass die Negerbevölkerung von Zentralafrika gegenüber der tropischen Schlafkrankheit weniger resistent ist als die eingewanderten Europüer ? Es wäre vielmebr zu erwarten dass die Selektion bei der schwarzen Bevölkerung eine erhöhte Immunität begünstigt hätte.

HALDANE. 1. I agree with Dr. Montalenti's project. Another possibility is that (by analogy with the advantage possessed by vermilion

Drosophila on media deficient in tryptophan) microcythemic heterozygotes may be at an advantage on diets deficient in iron or other substances, thus leading to anacmia.

2. Perhaps the theory that most diseases evolve into symbioses is somewhat panglossist. I doubt if it occurs as a general rule, though it may do so. The position for the original host is however best.

Index